Psychology and Our Planet

Series Editor

John Fraser, New Knowledge Organization Ltd.
New York, NY, USA

The book series Psychology and Our Planet comprises edited volumes, monographs, and briefings that focus on the interaction between mental process and the environment. At present environmental, population, and conservation psychology represent broad areas of practice that tend to have a disaggregated publication record. While the work in past builds on a broad base of research, synthesis of these papers tends to only appear in service to a specific research publication, and more importantly, is seldom synthesized in ways that are useful to academic training and policy advancement. The series seeks to redress that gap by providing a forum and meeting place for psychologists from across disciplines to advance the exchange ideas, and, where appropriate, provide opportunities for collaboration. The aim of the series is to publish books on the many dimensions of how people conceive themselves within the biological and cultural systems that shape the world, and to expand the full range of human relationships to the conditions that have created the world we live in today, and the decisions that will guide anthropogenic impacts on the planet's future.

Topics covered in this series include:

- Synergies between human mental health and the ecology of the natural environment, and the psychological consequences of high population density.
- The role of mental processes and human behavior in advancing a thriving biosphere on which all life depends.
- The understanding of reciprocal relationships well-being related to the design of built spaces, landscapes and natural environments.
- The reciprocal relationships between environmental conditions, be that the natural or built world, and the shaping of mental process.
- The synergies between human mental health and ecological systems.
- The psychological consequences of high population density of the environment in mental development and mental health.

Psychology and Our Planet welcomes book proposals representing the broad interest of the interdisciplinary and international focus of the series, and is particularly excited to receive proposals that address aspects of systems and habits that, at scale, may be able to describe and reverse the negative sequelae that are now placing people and populations in harm's way.

Christian A. Klöckner • Erica Löfström

Disruptive Environmental Communication

Springer
Knology

Christian A. Klöckner
Department of Psychology
Norwegian University of Science and Technology
Trondheim, Norway

Erica Löfström
Department of Psychology
Norwegian University of Science and Technology
Trondheim, Norway

ISSN 2662-1916 ISSN 2662-1924 (electronic)
Psychology and Our Planet
ISBN 978-3-031-17164-2 ISBN 978-3-031-17165-9 (eBook)
https://doi.org/10.1007/978-3-031-17165-9

This Springer imprint is published by the registered company Springer Nature Switzerland AG
The registered company address is: Gewerbestrasse 11, 6330 Cham, Switzerland

Preface

This book is the result of several years of discussion of a wild idea, first born at a workshop about radically new approaches to sustainable futures that we both attended in 2016. We discovered that we both were feeling more and more uncomfortable with the mainstream approaches to dealing with environmental problems within behavioral sciences and started questioning some of the predominant mantras like "you must not upset people with your communication campaigns" or "it is best to sneak in gradual changes in people's behavior without them even noticing." While we acknowledge the arguments for these approaches and their value in some contexts, we deeply question the assumption that they are the only viable pathway to change. We would even go so far as to say that—facing the urgency of societal transition and the extremely limited time spans we have to address problems like climate change, biodiversity loss, resource overuse, and waste—there is a need to explore more radical instruments for stimulating discussion and change. At some point, the concept of disruptive communication, which this book is presenting, was conceived.

During the years that followed that workshop, we presented the idea in workshops, conferences, and meetings, and we invited colleagues of us to explore and reflect on different aspects of the concept with us. This book represents the state of these discussions so far. In the nine chapters of this book, we introduce the concept, explore the psychological mechanisms that could make it attractive, discuss the thin line it is necessarily balancing on, and present examples of disruptive communication from the world of art, games, and design. We also reflect on what disruption means in a global context and the implications for intercultural projects. Finally, we present some first insights from our research project Nature in Your Face, which is a first larger attempt to bring the concept to life in four cases in Norway. We are deeply grateful for the contributions of our colleagues Berend Barkela, Kristoffer Fjællingsdal, Maja Grünzner, Amanda E. Lai, Teresa Gil López, Isabel Richter, and Laura K. Sommer. Their thoughts, ideas, and critical reflections have shaped this book, which has become an unusual hybrid between a monography and an edited volume with at least one of us involved in every chapter and where our ideas are challenged by our colleagues. We acknowledge our colleagues by listing their names of these contributors as chapter authors.

Since we started this book more than two years ago, the world has changed more rapidly than we could have imagined. First, a pandemic substantially disrupted our lives and with us the lives of billions of people around the world. Even more recently, while we are finalizing the book, a war in the middle of Europe is changing the political landscape more rapidly than we could have imagined just some months ago. While these events, on the one hand, show how powerful disruptive events can be to change people's lives and lifestyles, it also raised the question if promoting disruptive communication is ethical in times where massive disruptions interfere with our lives already. We argue that our concept is fundamentally different from just causing disruption and chaos, but rather is a tool to stimulate discussion, negotiation, co-construction, and empowerment of people. Designing the disruptive elements in a way that they are both intrusive and acceptable is the first challenge which we reflect on in this book, but also the follow up of the first disruptive experience with a structured process of dialogue, creation, and creativity. Understood in this way, disruptive communication might even be the opposite to the large disruptive events we see around us. It is possible that implementing the full cycle of disruptive communication as outlined in this book may even contribute to the revival of democratic ideals, a notion under threat in many countries around the world. We hope it does.

Trondheim, Norway

Christian A. Klöckner
Erica Löfström

Contents

The original version of this book was revised. A correction is available at https://doi.org/10.1007/978-3-031-17165-9_10

Chapter 1
What Is Disruptive Communication, and Why Might It Be Necessary?

Christian A. Klöckner and Erica Löfström

1.1 Introduction

The list of sustainability issues our societies are confronted with is long, and human behavior is at the core of what is causing many of these problems (e.g., Pachauri et al., 2014; Brondizio et al., 2019). However, even if almost all countries of the world in an unprecedented process committed to the United Nations Sustainable Development Goals in 2015, human life on planet Earth is facing severe challenges, and the rate of improvement seems to be far too slow to reach the goals in time. This raises the question of why people in all kinds of societal functions (including decision-makers and ordinary citizens) fail to engage more in the changes needed. Undoubtedly, behavioral scientists in disciplines such as psychology, sociology, human geography, behavioral economics, or product design have developed profound knowledge about drivers of human behavior and barriers toward change (e.g., Gifford, 2011). They have developed strategies to change people's behavior and have – together with policymakers and other actors – implemented small- and large-scale programs to tackle these challenges. What seems to characterize a vast majority of these strategies is the implicit rule that citizen engagement can only be achieved if citizens are not upset by the implemented intervention instruments and subsequent changes. An underlying assumption seems to be that resistance is to be avoided and that changes that do not require people's active engagement are preferred. This has, for example, lead to the nudging approach (Demarque et al., 2015; Thaler & Sunstein, 2009), which aims to design choice situations in a way that makes people take the "right" choices without the burden of thinking about their decision. This approach has received strong interest among policymakers, apparently not the least because it promised less resistance among citizens (and voters).

The success of behavior science-informed programs has varied a great deal, some resenting impressive results (Goldstein et al., 2008; Schultz et al., 2015), while others yield only small effects, also depending on the kind of intervention

C. A. Klöckner, E. Löfström, *Disruptive Environmental Communication*, Psychology and Our Planet, https://doi.org/10.1007/978-3-031-17165-9_1

(Leeming et al., 1993). On a large scale, it is safe to conclude that societies do not manage to initiate fundamental change processes with the governance and engagement methods that are in use today. This calls for exploring new and potentially more disruptive forms of intervention. While we accept that contemporary behavioral science interventions (including nudging) have proven that they can be effective under certain conditions and thus have their place in pushing the societal transitions necessary, we will, in this chapter and in this book, challenge the assumption that this will be enough to achieve the changes in the limited timeframe we have. In the following section, we will build this argument by briefly analyzing the status of the planet, outlining the emergency. Then we will shift to an analysis of the societal responses, before diving into the reasons for why we are not more radically address what is needed to reach the transformation goals. Afterward, we will leave the realm of behavioral science to study how other disciplines look at transition points, transformative changes, disruptions, and societal innovation. This prepares the introduction of the core concept of this book, namely, disruptive communication. This concept will be explored in the following sections before the remaining chapters of the book will be introduced in relation to the disruptive communication concept.

1.2 A Planet in Distress

The signals that the capacity of our planet is at or beyond its limits are plenty. Maybe most prominent in the media, climate change has received the most attention in recent years. The scenarios the Intergovernmental Panel on Climate Change (IPCC) draw in their most recent fifth assessment report are dire (Pachauri et al., 2014): global warming is unequivocal and human influence on the climate system is evident. Manifestations of global climate change are already happening with droughts, wildfires, heat waves, extreme weather events, etc., and all the effects they have on ecosystems and human life are severe (e.g., food security, property, livelihoods, and health). Even if global climate change is unequivocal with the amount of CO_2 in the atmosphere already, decisive action to mitigate its effects is necessary. If the current trajectories remain unchanged, future impacts of global warming are predicted to be magnitudes more immense than what we already experience, including the risk of passing thresholds of unreversible changes and tipping points, according to IPCC reports. The science behind the report indicates that there are viable pathways to limit global warming to under 2 °C, but these require substantial reductions in CO_2 emission within the coming decades and a drastic reorientation of societies already until 2030. Policymakers in many countries have responded to this call with ambitious reduction targets (e.g., the green deal of the European Commission, outlining an action plan to reach ambitious climate goals and become the "first climate-neutral continent," European Commission (2019)), also pushed by the massive mobilization of the Fridays for Future movement in 2018 and 2019. It remains a significant challenge though to implement this plan in a Europe characterized by contradicting interests and with populations where political polarities are on the rise, also leading to stronger contrasts between climate change skeptics and promoters of climate action (Poortinga et al., 2019). In spring and summer 2022, climate actions in the EU have again come under pressure of measures counteracting the energy crisis caused by the war in Ukraine.

However, climate change is not the only human-made environmental challenge. In 2019, the Intergovernmental Science-Policy Platform on Biodiversity and Ecosystem Services (IPBES) published their first assessment report on biodiversity and ecosystem services (Brondizio et al., 2019). They conclude that nature is deteriorating on a global scale and that this impairs the services nature provides for humans. About every fourth animal or plant species (more than one million in total) are expected to face extinction within the next decades if no actions are taken, among them species of high importance for human survival. This profound change of ecosystems is caused by human activity, including extensive use of land and sea habitats, pollution, direct exploitation of organisms, and human contribution to spreading alien species to habitats. The report concludes that the underlying causes are societal values and behaviors, production and consumption patterns, and human population dynamics, factors which again are within the realm of behavioral science. The report further states that sustainability, with respect to biodiversity, is impossible to reach within the current trajectories and requires transformative changes across economic, social, political, and technological factors.

A third major environmental issue is the overuse of limited resources. This applies, for example, to the extensive use of freshwater (Bogardi et al., 2013), phosphorus as a key ingredient in fertilizer production (Ragnarsdóttir et al., 2011), or metals and rare earths providing the backbone of infrastructure and electronic devices (Sala et al., 2020). Even if environmental challenges do not stop with these three major domains, they nonetheless represent the potentially most pressing issues, and a recurring notion in all reports is that a sustainable future is only possible if lifestyles, production, and consumption patterns are changed fundamentally and within a very short timeframe.

1.3 Societies Fixing Details but Ignoring the Fundamental Change Needed

Given the state the planet is in and the possible consequences it has for the future of humanity, immediate action should be self-evident. And in fact, there are actions taken both in the political and private sphere. We already mentioned the European Commission's Green Deal (European Commission, 2019), which is probably one of the most ambitious political programs proposed by majority policymakers. Also, citizen initiatives of smaller and larger scale addressing environmental issues emerge in growing numbers (see, e.g., www.rescoop.eu for an overview of citizen-driven energy initiatives) and understanding why and how individuals join such initiatives is growing (Tosun & Schoenefeld, 2017). In the production sector, a shift of perspective can be recognized, and environmental impacts have become a relevant decision dimension, often, however, more as a response to stakeholder pressure than genuine environmental concern (Littlewood et al., 2018). There is also a large variation in how especially CO_2 intensive industries address environmental issues such as climate change in their strategies, ranging from denial and active opposition to policies to acceptance and exploration of alternative future business models (Skjærseth & Skodvin, 2018). Interestingly, cities rather than regional, national, or multinational actors have emerged to be drivers of environmental, especially climate action (Hughes et al., 2018). This is both a necessity and an opportunity, as more than half of the world's population lives in urban settings, and cities contribute largely to emissions (Bai et al., 2018). On the other hand, cities often have larger flexibility to find local arrangements.

However, despite all these promising initiatives on all societal levels, the IPCC and IPBES reports tell us clearly that we are not on track to reaching climate or biodiversity targets (Brondizio et al., 2019; Masson-Delmotte et al., 2019). The degree of change is neither fast nor deep enough to make sure that by 2030 the foundations for a sustainable society are in place. Our hypothesis is that this is at least partly due to that these initiatives overwhelmingly focus on fixing single symptoms rather than addressing the underlying structural causes of overconsumption and environmentally detrimental lifestyles. Examples are the electrification of car traffic, the insulation of large houses, or the electrification of oil production platforms. We like to be clear here, all these measures have a large potential to reduce emission levels in themselves, and we have no intention to devalue them, but they do not question the acceptance of the underlying need for individual car traffic, for large homes, or for the production of oil. Within an economic and societal system that for many people in the world has produced a stable increase in living standards, the attempts to fix the resulting environmental problems appear to be restricted to incremental and rather selective measures.

1.4 Why Are We Avoiding Shaking the Tree?

In light of what just has been said, it is an interesting question of why social scientists are avoiding being more confrontative in their communication and intervention strategies. In the literature, there are, in fact, some good reasons for avoiding confrontation brought forward. Maybe the most prominent argument is that interfering with peoples' lives will cause what psychologists refer to as reactance (Brehm & Brehm, 2013). In simple terms, reactance theory states that people often protect the freedom they have to choose between different behaviors. If an intervention reduces peoples' freedom of choice, they actively attempt to reinstate this freedom by fighting against the interventions or sabotaging their implementation. The more important the impaired behavior is for a person, the stronger the feeling of reactance. In a recent series of studies, Kavvouris et al. (2020) show that normative appeals to be environmentally friendly (recycle electronic devices) may lead to reactance and reduced intentions to act. In a first study, they found that especially injunctive norm appeals ("a large majority think that everyone should recycle their electronics") increased the perceived threat of behavioral freedom, as compared to descriptive norm appeals ("a large majority recycle their electronics"). This feeling of a threat then induced counterarguing. In the second set of two sub-studies, they confirm this chain from normative appeals to the threat of behavioral freedom and counterarguing and extend it further to reduced intentions to act. The authors conclude that the implication for campaign design is that normative appeals should preferably be presented as descriptions of what other people do, not what you should do. In simple words, campaigns should avoid moralizing. Kurz et al. (2020) follow a related line of argument in their paper about the impact of highly moralized behavior of minorities (e.g., vegans, all-year bicyclists) on the development of the behavior in the majority. They assume that the visibility of these groups and the high moral charge of their behavior might rather hinder than foster a behavioral shift of the large majority because the majority fears that they have to shift behavior drastically to be a member of these groups. By the high moral standards, the minority groups also threaten the way of life of the majority to the degree that can cause reactant behavior.

Also in the political sphere, environmental measures that have the potential to provoke reactance among the public tend to be avoided. Pull measures (e.g., subsidies for electric cars) are usually preferred over push measures (e.g., increasing taxes on gasoline), which is also mirrored in research indicating that pull measures usually receive better public support than push measures (Drews & Van den Bergh, 2016). In light of reactance theory, this can be explained by pull measures opening up for more choices, whereas push measures reduce the choice. In an analysis of European policy documents (Klöckner et al., 2018), the researchers concluded that the main underlying assumptions about how citizens make decisions are that they need to be informed about better alternatives and that the more environmentally friendly alternatives need to be made economically competitive. This goes nicely hand in hand with the popularity rather unobtrusive intervention measures such as nudging have among policymakers. In the nudging approach, the freedom of choice is explicitly maintained (as no behavior is banned), but the behavioral alternatives preferred by the agency implementing the intervention are made more accessible or easier to implement (Demarque et al., 2015). An example is default settings for 100% regenerative energy when a new tariff is signed by a customer of an electricity provider (Ebeling & Lotz, 2015), where a massive increase in green tariffs can be seen in an opt-out version (you have to say no to green electricity) as compared to an opt-in version of the website. What characterizes nudging approaches is that they either go completely unnoticed by the people making the choice in question (and thus do not cause reactance), or they are perceived as not threatening the freedom of choice. Because of its unobtrusive nature, the ethical background of nudging and choice architecture has recently moved more into the focus of the discussion. Sunstein (2015) concludes that nudging is not unethical but that the justification for implementing the nudges needs to be made by the authorities doing it.

Another line of argument that has gained some popularity in behavioral science is the approach to avoid moralizing and rather make environmental behavior fun, entertaining or a social event. In other words, connect environmental action to positive rather than negative emotional experiences (Cooke & Fielding, 2010). Stoknes (2014) argues that communicating about climate change in dystopic, catastrophic images, and narratives weakens people's motivation to act against it. Furthermore, he states that the predominant narratives are of loss and sacrifice (give up privileges as flying), which triggers the human tendency to go a long way to avoid losses (Kahneman & Tversky, 1982). In conclusion, Stoknes recommends focusing climate communication on positive messages, for example, by highlighting the positive effects of climate actions in other domains such as health (e.g., when promoting bicycling). Gamification of environmental actions is a trend related to connecting environmental action to positive experiences, to make it fun to do (Morganti et al., 2017). The assumption behind this approach is that connecting an environmental behavior such as energy-saving to game elements (e.g., a competition against other "players") creates an additional and potentially stronger motivation to perform them.

However, research shows also that more confrontative and disruptive strategies might be valuable to achieve changes in environmental values, attitudes, and, not the least, behaviors. Grube et al. (1994) presented, for example, an application of belief-system theory and proposed a self-confrontation strategy to make inconsistencies between values a person holds and their behavior visible. These ideas are essentially related to dissonance theory (Festinger, 1962), which assumes that people experience a state of cognitive tension when holding contradictory values or beliefs or when their behavior does not match their values. People would then aim to gain consistency in their cognitive system by changing one or several elements of the system. Thus, if a person embraces environmental values but acts against them, pointing that mismatch out to the person induces an unpleasant feeling of cognitive dissonance. This can be resolved by either changing behavior to be in line with the person's values, by changing the values so that they are in line with the behavior, or by defining situational conditions in a way that externalizes the source of the mismatch ("it is not my fault that I cannot be more environmentally friendly, I was forced by the circumstances"). Based on this theory, making a mismatch between peoples' values and their actions visible could be a motivator for behavior change. However, this requires (a) that the person holds pro-environmental values, to begin with, (b) that these values are more resilient to change than the behavior, and (c) that the person considers the values important and relevant enough in the given situation to care about the mismatch. In an experimental study, Rubens et al. (2013) found that an intervention strategy based on confronting people who signed a petition to reduce the use of plastic bags in supermarkets with their previous behavior (using plastic bags) was less effective than asking people for commitment to avoid plastic bags. They interpret this finding as an indication for that hypocrisy techniques (techniques outlining mismatches between stated values and behavior) often rather lead to an adjustment of the values than the behaviour.

Thus, based on the brief review of some relevant literature presented in this section, it is not surprising that behavioral science has moved strongly into the direction of not upsetting citizens when intervening to change environmental behavior. Interventions should avoid moralizing; they should not lead to reactance, maybe even go totally unnoticed; in the best case, people should enjoy environmental activities or their co-benefits. We like to be very clear that we do not reject these assumptions. All these recommendations have their place and good reasons. What we like to challenge, though, is the assumption that we can achieve the fundamental change of societies necessary exclusively with these techniques. The techniques tend to improve environmental footprints gradually; they do not question the bigger picture. Thus, we will, in the remainder of this chapter, argue for exploring the concept of disruptive communication as an alternative or supplement to the established techniques together with us.

1.5 Characteristics of Transformative Change Processes

While leaving the domain of behavioral science for a moment, it is interesting to see how other disciplines conceptualize drivers of fundamental societal change theoretically. In the area of sustainable transition research, some interesting work and reflections have been published that also shed some light on the conditions of transformative change. In a highly influential article, Geels (2010) analyzes sustainable transitions from seven different ontological approaches: rational choice, evolution, structuralism, constructivism, functionalism, conflict and power struggle, and relationism. What is striking in this analysis is that most ontologies predict that systems are attracted to stability and usually favor only incremental changes within the established system. However, under certain conditions (e.g., external shocks or impulses, change in higher-order cognitive frames), these systems destabilize, transform, and reorganize into new states of stability.

In organizational management research, the assumption that organizations oscillate between phases of stability with little change and phases of radical change is also rather common (Anderson & Anderson, 2010). When an organization or a company is new, it quickly develops and typically reaches a plateau of success at some point, where further development comes to a hold. After that plateau, the success of the organization first slowly then with more speed deteriorates. A stage of instability and chaos is entered often triggered by external events, where several "wake-up calls" might succeed in bringing the organization temporarily back on track. However, sustainable change is only achieved if, at some point, the mindset of the organization shifts fundamentally. We see this description of organizational development as a fitting analogy for the development of our societies: after a period of fast development during the industrial revolution, which was literally fuelled by coal and oil, the development reached a relatively stable plateau, from which the state of the societies started slowly to deteriorate. At the moment, we seem to be in a stage of repeated wake-up calls and trials to fix acute problems. However, the fundamental mind shift that would give birth to the next societal model is still pending.

Though not a scientific source but rather a "learning from the practice of change management" approach, Bob Tipton, the leader of a consultancy, formulates nine interesting stages of transformational change (Tipton, 2012), which can help to structure thoughts about change processes. His stages have a certain overlap with the model of change outlined above (Anderson & Anderson, 2010). According to Tipton's model, transformational change starts with the status quo, which in itself is usually perceived as comforting. However, either from the inside or (more often) from the outside, a change process is started, the comfortable status quo gets disrupted. The second stage of change is denial: in other words, ignoring the need for change and clinging to the status quo. In the third stage, members of the system openly resist; they get angry; they work against change. This stage has much in common with the reactance concept introduced above. The fourth stage, pleading, is characterized by wishful thinking to make the problem go away, whereas the fifth stage is despair or skepticism, where the choice has to be made for giving up or starting something new. Up to this point, the change process is characterized by grief about what is to be lost. If at the low point (despair) a decision is reached to try something new, a new process of growth will replace the grief stage. The first step of this growth process is – according to Tipton – tolerance. This stage is characterized by growing acceptance of the need that something needs to happen, and resistance fades, but there is still some negativity left. This can still lead to people retreating to earlier stages. In the next stage, acceptance of the necessity of change is established, which in the next stage transforms into an agreement, thus in active and positive support. The final stage, advocacy, is characterized by people themselves promoting the change. From the perspective of disruptive communication, this practice-based model is interesting for two reasons: (1) the necessity of an organizational change is often triggered by an external disruption which is first not accepted within the organization, and (2) Tipton claims that he tries to get organizations as soon as possible to the "despair" stage by disrupting their denial, resistance, and pleading stages actively. Furthermore, the stage model seems to match the stages of responses to climate change quite well: from being comfortable with the status quo, for many people, news about global climate change and necessary measures against it first led to denial, and then resistance or wishful thinking, which eventually reached a stage of despair and skepticism to that a viable solution is available.

1.6 Radical Innovation Instead of Incremental Change

We have all heard it, that necessity is the mother of invention or innovation. If so, how can this help us in combatting the threat of climate change? Can we stimulate radical innovation (RI) to combat and/or mitigate climate change? Climate scientists warn that global warming will increase the frequency and intensity of extreme weather events (Pachauri et al., 2014), and we have indeed seen examples of this happening over the last few years with floods, droughts, and tropical heatwaves, et cetera. However, the problems emerging with climate change are still more or less hidden, especially in the wealthier parts of the world, and we can still pretty much continue to live our lives the way we are used to most of the time. This means that we do not necessarily experience the urgency of changing into more sustainable practices before it is too late. Incremental change and incremental innovation have indeed contributed to making products and services per se more efficient, but consumption levels are still increasing worldwide (Ugursal, 2014). As an example, refrigerators have become less energy-demanding for keeping food cold (Boardman, 2004), but if the refrigerators also become *bigger and capable of storing more food*, the positive effects are eaten up. Thus, incremental change has not led to an overall lowering of consumption levels. The concept of sufficiency, which is closely linked to that of circular economy (European Commission, 2015) and the discussion on limits to growth or prosperity without growth (Jackson, 2017) introduced during later years, address this problem and has largely come to replace the more established concept of efficiency, but changes and policies are still very much incremental. Opposing these concepts is radical innovation. There are different definitions of radical innovation – sometimes also referred to as disruptive innovation – but most scholars agree that a radical *innovation needs to break with the existing* (system, process, service, or product) and, if successful, replace it with something new. As an example, Ettlie et al. (1984) argue that a "unique strategy and structural arrangements are necessary for radical innovation" (p. 683). Also, although the term innovation is often referred to with regards to technology or products, it is also used with regards to services and service design which "accentuates holistic, multi-disciplinary, and integrative characteristics" (Yang & Sung, 2016, p. 21). An emerging field in design – namely, speculative and critical design (SCD) – confronts traditional design practices by seeking to change the contemporary perception of products and services (Dunne & Raby, 2001, 2013). With regard to the threat of climate change, it is clear that we need to combat the overconsumption by radical innovation rather than continue on the current path of incremental change. The fact that consumption levels are increasing instead of decreasing while these noninvasive engineering techniques have been extensively used calls for alternative routes. We suggest instead a path to stimulate radical innovation using the urgency as a disruptive element to engage and stimulate discussions on how we can do things radically different at a systemic level. This offers a way forward into addressing climate change. We need to stimulate for radical innovation but also make sure that the creative solutions actually reach politicians and policymakers so they can be implemented.

1.7 The Concept of Disruptive Communication

So let us recapitulate the argument that we have built up in this chapter so far: we are in a situation where societies, as they work today, exceed the planetary boundaries by magnitudes and put the survival of the human race as we know it at risk. Our societies have so far responded to this threat by addressing single symptoms with incremental changes, which basically stabilize the status quo. In behavioral science, a general tendency seems to be avoiding upsetting citizens and rather try to steer their behavior by subtle nudges, by choice architecture, or by highlighting benefits of behavior change not related to environmental issues (such as health, comfort, or enjoyment). In other disciplines, the nonlinearity of transformative change processes appears to be more accepted, and indications can be found that a more radical disruption of the status quo may be a necessary precondition for societal transformations and radical innovation. However, such transformational processes can be painful and likely include a stage of grief. This feeling of grief in relation to climate change and the necessary transitions has been described by several authors (Cunsolo & Ellis, 2018; Westoby & McNamara, 2019).

This leads us to question the predominant mantra in the behavioral sciences that upsetting people, pushing people out of their comfort zone, is a no-go for environmental campaigns. On the contrary, we advocate disruptive communication techniques as an alternative avenue of behavioral interventions. However, disruption is not a goal in itself; it is necessarily linked to follow-up processes harvesting the emotional activation a disruption of everyday life causes. However, before we dive deeper into a discussion of what disruptive communication is and what it should achieve, we present our working definition:

> Disruptive communication is communication in a broad sense (including language-based and non-language-based means of communication) that aims at interrupting everyday life, which stops people from continuing their business-as-usual, that confronts accepted assumptions and that provokes emotional engagement (positively or negatively). Disruptive communication interferes with normal ways of living, aiming at stimulating processes of radical change.

Figure 1.1 displays some of the key characteristics of disruptive communication as we see it in more detail. The main aim of disruptive communication is to interrupt citizens, policymakers, businesses, organizations, etc. in their standard way of behaving. Something unexpected happens that draws attention and makes the established behavioral patterns unsuitable. Let us take a hypothetical example: imagine a city that has a serious problem with excessive car travel. One morning, one of the main streets into the city center is suddenly occupied by an artistic performance theater staging mobility solutions of the future in a creative way (e.g., by "riding" large mechanical human-powered structures or "sailing" the street in "solar storm driven" street floats). You cannot drive to your workplace the way you are used to. This provokes a response from you: you might get upset, you might be afraid to come too late for work, you might get angry, or you might be amused by this interruption of your daily routines. You might feel so angry that you leave your car and shout at the actors. Whatever is your reaction, the point with this communication attempt is to push you out of your comfort zone and provide you with emotional energy. You care about the topic, even if this at first might be because the event questions your way of living. The art of this approach lies in finding the balance between stopping people in their tracks and, at the same time, impressing them with the unconventional means taken into practice. So, for us, disruptive communication is

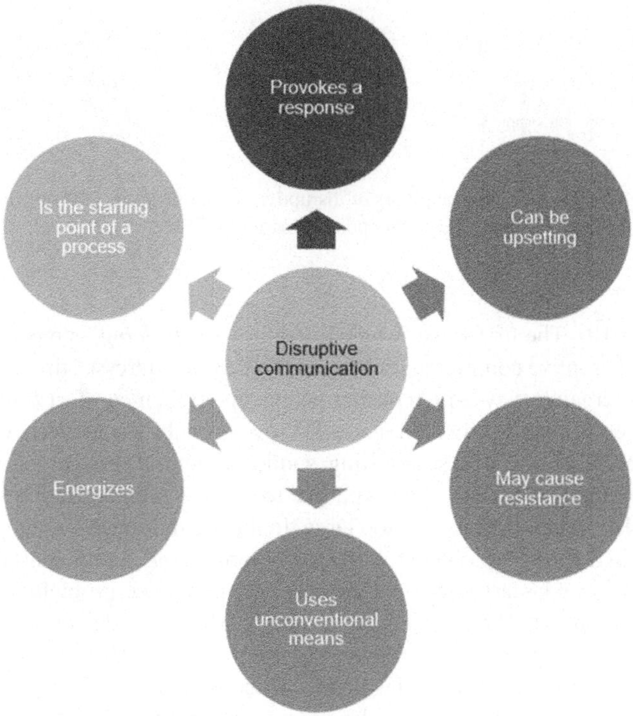

Fig. 1.1 What is disruptive communication?

not simply annoying people; it is creating situations where there is a tension between being challenged and provoked but also fascinated by what is happening. Most importantly, disruptive communication needs to have a strategy for following up on the initial response people might show. It is OK and expected that people get angry at first; disruptive communication embraces this response. However, people need to be taken seriously and invited into a process where this initial emotional activation is channelled into a creative, open space for developing radically new solutions. So, in our hypothetical example, the street theater could invite the citizens who were stuck in their tracks to a participatory planning process of the mobility of the future in the city. In this process, all relevant actors (citizens, planners, policymakers, interest groups, representatives of minorities, etc.) should be represented. What distinguishes our strategy from "ordinary" participatory planning processes is, on the one hand, the initial event which energizes the process and is supposed to bind the actors together and, on the other hand, the openness for more radical ideas in the structured follow-up.

As we indicated in the last paragraph, disruptive communication is more than provocation; it also includes a strategy for the follow-up. We conceive this follow-up process to be a three-stage process, which is displayed in Fig. 1.2 (Löfström &

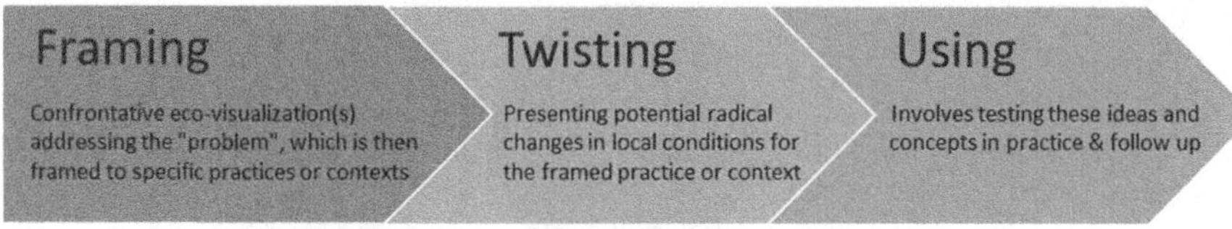

Fig. 1.2 The three methodological pillars of disruptive communication. (Taken from Löfström & Klöckner, 2019, page 63, reproduced with permission by ECEEE)

Klöckner, 2019). The first stage, which we refer to as *framing*, is mostly characterized by the disruptive communication event. This event addresses the environmental problem in a creative way, which pushes people out of their comfort zone and leaves them in an emotionally activated state. The event could use art, eco-visualization (Löfström & Svanæs, 2017), storytelling, sudden structural changes, or the like. The framing stage is also meant to help people to put the problem into a manageable context so that it does not become too large. In the second stage, *twisting*, people are confronted in a workshop context with radical, provoking, and potentially not realizable ideas to kick-start a thought process for innovative problem-solving. This approach is related to the ideas of provotyping, which is a design tradition where extreme, unusable, provocative prototypes of new designs are developed to stimulate the interaction with lay co-designers (Mogensen, 1992). Everybody is aware that the provotypes are by no means the end product, but their extreme nature pushes creativity and the borders of what is considered feasible. Thereby, more innovative solutions are found. However, the last stage, *using*, is also important to make sure the developed ideas from the twisting stage are actually implemented in real life (even if so in a modified version). This makes it important that all relevant stakeholders participate in this process, including the "decision-makers." It also makes a clear focus on balancing out power differences between stakeholders and principles of fairness essential for the process.

Actually, some actors in the environmental domain already use some of the principles of disruptive communication outlined here. At least part of the success Greta Thunberg and the Fridays for Future movement (https://fridaysforfuture.org/) have with putting climate change back on the agenda is the disruptive communication techniques they use. Greta Thunberg speaks in a way to powerful policymakers that one hardly expects from an adolescent. This simply interrupts the business-as-usual of climate policymaking. Some interventions of Extinction Rebellion (https://extinctionrebellion.uk/) also have clear indications of the disruptive communication elements, for example, when street theater acts are involved. What we see lacking in these examples, though, is the strategy for following up the engagement caused by the disruptive communication elements.

1.8 Some Thoughts on COVID-19 in Relation to the Topic of This Book

In 2020, almost all countries of the world experienced one of the most substantial disruptions they have experienced in modern times with the COVID-19 pandemic and the measures taken to confine the infection rates to a manageable level. For many months, far-reaching restrictions of people's behavior have been implemented by governments around the world from social distancing over lockdowns, travel restrictions, and cancellations of large events to changes in job arrangements. The speed and extent of the taken measures have demonstrated that societies, in fact, are able to respond quickly to existential threats, implement measures with large impacts, and – at least for some time – unite against the threat. As a by-product, the COVID-19 actions also demonstrated that it is practically possible to reduce pollution and CO_2 emissions overnight by up to 25% (Le Quéré et al., 2020), with tremendous costs and sacrifices for societies and individuals, though.

This is interesting also from the perspective of this book, as it clearly shows that actions are possible when the disruption is immediate and large enough, that they are supported by citizens to a large extent, but that this support is fragile and depends on how it is followed up, who benefits, and who carries the burden. Questions of fairness and justice will be of extraordinary importance in such processes, and in the COVID-19 crisis, lessons can be learned for how to manage large-scale transitions of societies.

1.9 Structure of the Book

After this chapter laid the foundations of the concept of disruptive communication, arguing for why disrupting business-as-usual can be an important tool in triggering societal change, the following chapters will apply this concept in different examples and explore more in-depth what the effects of the disruptive elements can be. For this purpose, we invited a number of co-writers for these chapters with whom we discussed the concept and asked to explore it within their area of expertise together with us. However, before we dive into these examples, Chap. 2 analyzes what psychology can teach us about change and its relation to discontinuities.

Chapter 3 explores the role of art as a disturbing element. Many artists respond to environmental issues in their practice and create artwork that is inspired by these issues. Often such artworks are uncomfortable to watch and interfere with daily life, at the same time as attending them is a rewarding experience for the spectator. Based on findings from the CLIMART project (www.climart.info), the chapter presents some conclusions about under which circumstances art can be considered an element of disruptive communication.

In Chap. 4, we discuss in detail the thin line that disruptive communication (as enacted by for example climate protest movements) needs to follow to not be disregarded. This chapter shows how utilizing elements of art and happening in the protest forms might contribute to receive more support in the general population and stimulate a deeper reflection of the aims of the protest movement.

In Chap. 5, we explore the potential of games for being disruptive elements that shake people up. Initially, it might be counterintuitive to analyze games, which here include all forms of digital and physical games designed to entertain people, under the umbrella of disruptive communication as they are usually rather contributing to escaping the disturbing real world for some hours. However, there is a growing number of games that focus on more than entertainment, and that even take the element of disturbing their players into use.

Chapter 6 takes a journey into the world of product design and explores a design strategy referred to as provotyping, namely, breaking out of the standard product design process which is conducted by "design professionals" and usually stays within the safety limits of what is commonly accepted. Provotyping is a design strategy taking citizens into the design process, actively encouraging the design of (often overly) extreme prototypes and by that challenging the status quo of what is possible and doable. As such, it clearly has features of disruptive communication, being a disruptive design strategy.

Chapter 7 takes a global perspective and explores how disruptive communication can be applied in non-Western contexts and how communication strategies might have to be adapted to different target groups. Further, the chapter will look at the role so-called perceptual future scenarios can play as a tool of communicating the need for urgent sustainable behavior change, especially in the developing world.

Chapter 8 explores if virtual reality and immersive virtual experiences may be a promising vehicle for realizing disruptive experiences that go beyond what can be implemented in the physical reality. With examples from flood prevention, the ideas are linked to climate change adaptation and mitigation.

Chapter 9, finally, connects the threats from the analyses conducted in Chaps. 3, 4, 5, 6, 7, and 8. It will showcase some first applications of the concept in the Norwegian action research project Nature in Your Face (www.niyf.no) and discuss lines for future research.

References

Anderson, L. A., & Anderson, D. (2010). *The change leader's roadmap: How to navigate your organization's transformation* (Vol. 384). Wiley.

Bai, X., Dawson, R. J., Ürge-Vorsatz, D., Delgado, G. C., Barau, A. S., Dhakal, S., et al. (2018). *Six research priorities for cities and climate change*. Nature Publishing Group.

Boardman, B. (2004). Achieving energy efficiency through product policy: The UK experience. *Environmental Science & Policy, 7*(3), 165–176.

Bogardi, J. J., Fekete, B. M., & Vörösmarty, C. J. (2013). Planetary boundaries revisited: A view through the 'water lens'. *Current Opinion in Environmental Sustainability, 5*(6), 581–589.

Brehm, S. S., & Brehm, J. W. (2013). *Psychological reactance: A theory of freedom and control*. Academic Press.

Brondizio, E., Settele, J., & Díaz, S. (2019). IPBES. *2019 Global assessment report on biodiversity and ecosystem services of the Intergovernmental Science-Policy Platform on Biodiversity and Ecosystem Services*. Bonn, Germany.

Cooke, A., & Fielding, K. (2010). Fun environmentalism!: Potential contributions of autonomy supportive psychology to sustainable lifestyles. *Management of Environmental Quality: An International Journal, 21*(2), 155–164.

Cunsolo, A., & Ellis, N. R. (2018). Ecological grief as a mental health response to climate change-related loss. *Nature Climate Change, 8*(4), 275.

Demarque, C., Charalambides, L., Hilton, D. J., & Waroquier, L. (2015). Nudging sustainable consumption: The use of descriptive norms to promote a minority behaviour in a realistic online shopping environment. *Journal of Environmental Psychology, 43*, 166–174.

Drews, S., & Van den Bergh, J. C. (2016). What explains public support for climate policies? A review of empirical and experimental studies. *Climate Policy, 16*(7), 855–876.

Dunne, A., & Raby, F. (2001). *Design noir: The secret life of electronic objects*: Springer Science & Business Media.

Dunne, A., & Raby, F. (2013). *Speculative everything: Design, fiction, and social dreaming*. MIT Press.

Ebeling, F., & Lotz, S. (2015). Domestic uptake of green energy promoted by opt-out tariffs. *Nature Climate Change, 5*(9), 868.

Ettlie, J. E., Bridges, W. P., & O'Keefe, R. D. (1984). Organization strategy and structural differences for radical versus incremental innovation. *Management Science, 30*(6), 682–695.

European Commission. (2015). Closing the loop–an EU action plan for the circular economy. *COM (2015) 614 final [Internet]*.

European Commission. (2019). The European Green Deal. *Communication from the Commission to the European Parliament, the European Council, the Council, the European Economic and Social Committee and the Committee of the Regions. Brussels*, 24.

Festinger, L. (1962). *A theory of cognitive dissonance* (Vol. 2). Stanford University Press.

Geels, F. W. (2010). Ontologies, socio-technical transitions (to sustainability), and the multi-level perspective. *Research Policy, 39*(4), 495–510.

Gifford, R. (2011). The dragons of inaction: Psychological barriers that limit climate change mitigation and adaptation. *American Psychologist, 66*(4), 290.

Goldstein, N. J., Cialdini, R. B., & Griskevicius, V. (2008). A room with a viewpoint: Using social norms to motivate environmental conservation in hotels. *Journal of Consumer Research, 35*(3), 472–482.

Grube, J. W., Mayton, D. M., & Ball-Rokeach, S. J. (1994). Inducing change in values, attitudes, and behaviours: Belief system theory and the method of value self-confrontation. *Journal of Social Issues, 50*(4), 153–173.

Hughes, S., Chu, E. K., & Mason, S. G. (2018). *Climate change in cities*. Springer.

Jackson, T. (2017). *Prosperity without growth: Foundations for the economy of tomorrow*. Routledge.

Kahneman, D., & Tversky, A. (1982). The psychology of preferences. *Scientific American, 246*(1), 160–173.

Kavvouris, C., Chrysochou, P., & Thøgersen, J. (2020). "Be careful what you say": The role of psychological reactance on the impact of pro-environmental normative appeals. *Journal of Business Research, 113*, 257–265.

Klöckner, C. A., Andres, J., Chebaeva, N., Dimitrova, E., Frieden, D., Koksvik, G., … Velte, D. (2018). *An analysis of the potential of advanced social science knowledge in policymaking*. Retrieved from Trondheim: https://www.echoes-project.eu/sites/echoes.drupal.pulsartecnalia.com/files/D%203.3%20final.pdf

Kurz, T., Prosser, A. M., Rabinovich, A., & O'Neill, S. (2020). Could vegans and lycra cyclists be bad for the planet? Theorizing the role of moralized minority practice identities in processes of societal-level change. *Journal of Social Issues, 76*(1), 86–100.

Le Quéré, C., Jackson, R. B., Jones, M. W., Smith, A. J., Abernethy, S., Andrew, R. M., et al. (2020). Temporary reduction in daily global CO 2 emissions during the COVID-19 forced confinement. *Nature Climate Change, 10*, 647–653.

Leeming, F. C., Dwyer, W. O., Porter, B. E., & Cobern, M. K. (1993). Outcome research in environmental education: A critical review. *The Journal of Environmental Education, 24*(4), 8–21.

Littlewood, D., Decelis, R., Hillenbrand, C., & Holt, D. (2018). Examining the drivers and outcomes of corporate commitment to climate change action in European high emitting industry. *Business Strategy and the Environment, 27*(8), 1437–1449.

Löfström, E., & Klöckner, C. A. (2019). *Nature in your face: Framing co-creative visioning.* Paper presented at the ECEEE summer school, Hyeres, France.

Löfström, E., & Svanæs, D. (2017). *Eco-visualization: An exploration of the concept and its practical implications.* Paper presented at the ECEEE, Nice.

Masson-Delmotte, V., Zhai, P., Pörtner, H.-O., Roberts, D., Skea, J., Shukla, P. R., ... Waterfield T. (2019). *Global warming of 1.5°C. An IPCC special report on the impacts of global warming of 1.5°C above pre-industrial levels and related global greenhouse gas emission pathways, in the context of strengthening the global response to the threat of climate change, sustainable development, and efforts to eradicate poverty.* Retrieved from Geneva, Switzerland

Mogensen, P. (1992). Towards a provotyping approach in systems development. *Scandinavian Journal of Information Systems, 4*(1), 5.

Morganti, L., Pallavicini, F., Cadel, E., Candelieri, A., Archetti, F., & Mantovani, F. (2017). Gaming for Earth: Serious games and gamification to engage consumers in pro-environmental behaviours for energy efficiency. *Energy Research & Social Science, 29*, 95–102.

Pachauri, R. K., Allen, M. R., Barros, V. R., Broome, J., Cramer, W., Christ, R., ... Dasgupta, P. (2014). *Climate change 2014: Synthesis report.* Contribution of Working Groups I, II and III to the fifth assessment report of the Intergovernmental Panel on Climate Change: IPCC.

Poortinga, W., Whitmarsh, L., Steg, L., Böhm, G., & Fisher, S. (2019). Climate change perceptions and their individual-level determinants: A cross-European analysis. *Global Environmental Change, 55*, 25–35.

Ragnarsdóttir, K. V., Sverdrup, H., & Koca, D. (2011). Challenging the planetary boundaries I: Basic principles of an integrated model for phosphorous supply dynamics and global population size. *Applied Geochemistry, 26*, S303–S306.

Rubens, L., Gosling, P., Bonaiuto, M., Brisbois, X., & Moch, A. (2013). *Being a hypocrite or committed while I am shopping? A comparison of the impact of two interventions on environmentally friendly behaviour*. Environment and behaviour.

Sala, S., Crenna, E., Secchi, M., & Sanyé-Mengual, E. (2020). Environmental sustainability of European production and consumption assessed against planetary boundaries. *Journal of Environmental Management, 269*, 110686.

Schultz, P. W., Estrada, M., Schmitt, J., Sokoloski, R., & Silva-Send, N. (2015). Using in-home displays to provide smart meter feedback about household electricity consumption: A randomized control trial comparing kilowatts, cost, and social norms. *Energy, 90*, 351–358.

Skjærseth, J. B., & Skodvin, T. (2018). *Climate change and the oil industry: Common problem, varying strategies.* Manchester University Press.

Stoknes, P. E. (2014). Rethinking climate communications and the "psychological climate paradox". *Energy Research & Social Science, 1*, 161–170.

Sunstein, C. R. (2015). *Nudging and choice architecture: Ethical considerations.* Yale Journal on Regulation.

Thaler, R. H., & Sunstein, C. R. (2009). *Nudge: Improving decisions about health, wealth, and happiness.* Penguin.

Tipton, B. (2012). *Nine stages of transformational change.* Retrieved from https://teamtipton.com/nine-stages-of-transformational-change/

Tosun, J., & Schoenefeld, J. J. (2017). Collective climate action and networked climate governance. *Wiley Interdisciplinary Reviews: Climate Change, 8*(1), e440.

Ugursal, V. I. (2014). Energy consumption, associated questions and some answers. *Applied Energy, 130*, 783–792.

Westoby, R., & McNamara, K. E. (2019). Fear, grief, hope and action. *Nature Climate Change, 9*(7), 500–501.

Yang, C.-F., & Sung, T.-J. (2016). Service design for social innovation through participatory action research. *International Journal of Design, 10*(1), 21–36.

Chapter 2
Business as Usual Forever? Psychological Mechanisms of Inaction and How Disruptive Communication Might Help

Christian A. Klöckner and Isabel Richter

2.1 Introduction

The concept of disruptive communication in the previous chapter argued for why we think it might help the necessary societal transitions to (also) disrupt established ways of doing. This chapter will dive deeper into some of the psychological mechanisms behind people's difficulties to implement fast and drastic changes. We will start with a more general analysis of how change might be perceived as a threat, triggering rather defensive instead of receptive mechanisms in many situations. Subsequently, we will elaborate on more in-depth mechanisms such as the status quo bias (which partly has its origin in specific decision-making heuristics), habits and routines, stabilizing social norms and values, and identities (Fig. 2.1). We then turn back to the interactions between the structural and the psychological level and study how lock-in situations are created. In the key part of this chapter, we will explore the potential of disruptive communication to rock these barriers to behavior change and present some hypotheses about by which mechanisms this might happen. We will also explain and discuss the stages of behavioral change and related implications for intervention design as well as how disruptive communication might help people to move between action stages. We will wrap this chapter up, shortly discussing risks and side effects of using disruptive communication to break respective mechanisms.

2.2 Life Change as a Source of Stress

Imagine the following situation: you are working in a company, and you are satisfied with your job. Your colleagues are nice, and you have a good relationship to your superiors. The service that your company offers is selling well, and you feel that you are able to contribute meaningfully to creating it. There are, however, some

C. A. Klöckner, E. Löfström, *Disruptive Environmental Communication*, Psychology and Our Planet, https://doi.org/10.1007/978-3-031-17165-9_2

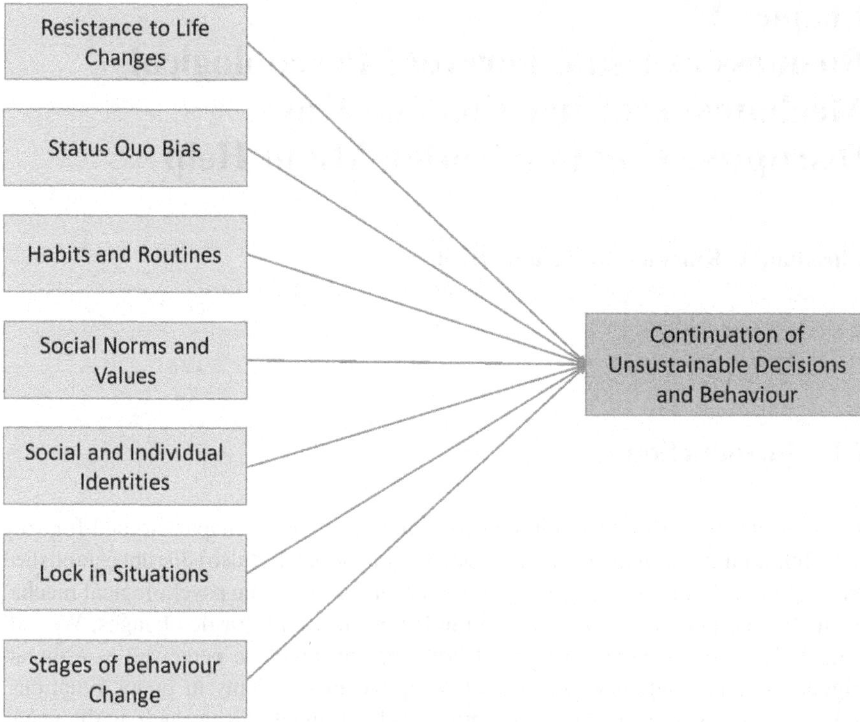

Fig. 2.1 Selected psychological mechanisms causing behavioral inertia

first signs indicating that the market might change in the future. Therefore, the top-level managers in your company who reside in a different country decide a substantial reorganization of the company: you will get new tasks in the company, might be moved to another department, and will be provided with a training for the new job. For most of us, such a scenario would cause stress and lead to a period of unrest; maybe we will even become unwell. The top-down change imposed by the managers is experienced as a threat toward what we are used to, who we are in the company, our position in the team, and established social relations. It challenges our competence and our knowledge; it makes our habits obsolete and requires that we invest a lot of energy in finding our way into the new situation. And this is "only" a change at the job level. The fundamental lifestyle changes required to address the ecological and social challenges our societies face are much larger. These changes span across all domains of our life: from work to family to free time. Not surprisingly, this scenario is perceived as threatening by many.

A substantial amount of literature has since the 1960s addressed the issue of how changes in peoples' lives contribute to stress, coining the concept of "life-change stress." Rahe (1975), for example, analyzed the available measurements for life-change stress for epidemiological studies and made some recommendations for how to best capture this concept. Central in his conclusions is that life-change stress may originate from several life domains, including health, work, home and family, the personal and social life, as well as the financial situation. If the changes are substantial in one or several of these domains and if people find it hard to deal with them, stress levels rise. If we take climate change as an example, mitigation but also adaptation will likely affect several domains and not all changes will be easy to implement for everyone. Stressful life changes have been linked to a number of psychological and physiological reactions and illnesses such as depression and anxiety (Mallinckrodt et al., 1989), reduced immune system (Irwin et al., 1990), or gastric cancer (Lehrer, 1980). Even if the relations between life-change stress and such severe outcomes are usually weak, it needs to be kept in mind that environmental crises like climate change or biodiversity loss have recently been associated with feelings of constant worry, anxiety, or insomnia in some people (Clayton, 2020; Marczak et al., 2021; Ogunbode et al., 2021; Panu, 2020), which might cause impairments of their daily lives over time and induce psychosomatic symptoms. This is particularly concerning as climate change engagement was recently driven primarily by adolescents in the Fridays for Future movement, indicating that the young generation is at risk. Therefore, communication strategies avoiding or at least buffering chronic negative emotional effects and stress which at the same time are effective in inducing change are necessary (Ojala, 2012; Pihkala, 2020).

In an influential paper, Cohen and Hoberman (1983) outline that positive life events and social support can be powerful buffers against the negative effects of stress caused by life changes. As a conclusion, we need to be careful about inducing changes in peoples' lives without additional positive experiences and clear structures to provide this buffer and support. In a significant study, Ojala (2012) found that most children and adolescents use several strategies to deal with the emotional distress caused by hearing about climate change: they de-emphasize the seriousness of climate change, distance themselves from the issue, trust in science or other people to solve the issue, or try to focus on the problem by taking actions against climate change. However, a significant group (about 5% of the studied children and adolescents) followed a strategy which Ojala refers to as hyperactivation, characterized by increased and persistent negative emotions, such as worry, resignation, or a feeling of powerlessness. In the group of young adults, this reaction was found in more than every tenth respondent. This again points toward the vulnerability of young adults for what Clayton (2020) describes as climate anxiety, also shown in a cross-national survey (Hickman et al., 2021). Pihkala (2020, 2022) points out the necessity to be aware of these effects and address them in environmental education and in pastoral care.

2.3 Status Quo Bias

In a well-known paper, Samuelson and Zeckhauser (1988) describe a phenomenon that is of high importance for understanding the inability of human societies to adjust to new conditions quickly enough, namely, the status quo bias. They argue that – unlike in economic textbooks – people make decisions in most situations with one alternative already implemented. If people decide to buy a new car or not, they often already own a car, for example. There is thus almost always the option to do "nothing" and leave things as they are. Interestingly, empirical findings as shown in Samuelson and Zeckhauser (1988) demonstrate that there is a bias in human decision-making favoring this "keeping the status-quo solution" over any kind of change, even under conditions where changing would be beneficial for the individual. That does not mean that change is impossible, but it means that when there is a choice between *acting* and *not acting*, *not acting* is perceived as more favorable than it should be and *acting* or changing is perceived as less favorable.

In their paper, Samuelson and Zeckhauser discuss a number of reasons for this effect to happen: first, they argue that a status quo bias might very well be rational decision-making, as the selection between two alternatives where one is already implemented needs to take into account potential transition costs, which might make a change toward a slightly better alternative unfavorable, nonetheless. Examples are manifold: public organizations, for example, change their software systems every couple of years after a tendering process, which has resulted in yet another competitor giving the apparently better offer. Often, the not directly assessable concomitant transition costs are neglected. To take a climate change-related example, switching the heating technology in a house from an oil-fired central heating to heat pumps is beneficial from an environmental perspective in most cases. However, the transition costs (not only in terms of investments to be made, but also in terms of disruption of everyday life) may be high. When the transition costs are unknown, this background for the status quo effect stands particularly strong (Samuelson & Zeckhauser, 1988).

Changes in our societies due to environmental challenges have a high degree of unknown transition costs, especially on the individual level, which then increases the tendency toward the status quo. We do, for example, not know how costs for different energy carriers will develop in the future and if and when investments in new heating technologies will pay off. Furthermore, once a person has decided for one alternative, this alternative is perceived as preferable. The mechanism behind this effect is that the uncertainty about the performance and outcomes of the chosen alternative is reduced, whereas potential alternatives are still equally uncertain (Samuelson & Zeckhauser, 1988). For heating technology choices, this means again that a technology we know and have relied on for many years has less uncertainty connected to it, than a new technology we do not know. This again leads to situations, where certain behaviors or technologies lock themselves in and become resistant to change. This will be analyzed more in Sect. 2.7.

We cannot always explain the status quo effect through rational choice principles. In their highly influential work, Tversky and Kahneman (1991) could show that people value losses and gains subjectively different. Losses seem to weigh more than gains of the same size. When a change of a status quo is related to potential losses, the gains from the alternative (which is not known from direct experience and more uncertain for the person) need to be higher than the losses to outweigh this cognitive bias. This effect has also been referred to as the endowment effect in the literature (Kahneman et al., 1991). This effect is particularly relevant if investments need to be made for environmental causes: using the heating technology example again, a situation where the upfront investment (framed as a loss) is to be made now and the potential gains (saving of energy costs in the future) are more uncertain requires a substantial imbalance in favor of the gains to become attractive.

Samuelson and Zeckhauser (1988) argue that the endowment effect is not the only explanation for a non-rational status quo bias. Another explanation could be the so-called anchoring heuristic. The anchoring heuristic states that people make decisions by taking the already established behavior as an anchoring point, which is then adjusted by new information coming in. However, this adjustment is frequently not strong enough to change perceptions completely in favor of the new behavior. Residues of the old behavior's initially favorable evaluation are drawing the new evaluation in the direction of the previously existing cognitive anchor (Tversky & Kahneman, 1974). Regarding the status quo effect, this heuristic anchoring of a judgment in relation to previous decisions will give the already established status quo a relative advantage in the new evaluation. Switching diet from a carbon-intensive meat-based diet to a more climate friendly plant-based diet might serve as an example here, where the initially favorable evaluation of the meat dishes may draw the evaluation of the vegetarian dishes down.

Samuelson and Zeckhauser (1988) name a number of other psychological biases and effects that might motivate the status quo bias, such as the sunk costs effect (Arkes & Blumer, 1985), regret avoidance (Kahneman & Tversky, 1982), cognitive dissonance (Festinger, 1962), or self-perception (Bem, 1972). The sunk cost effect refers to that people tend to stick to a decision they made before if they already invested a substantial amount of money, time, or effort into it, even if they realize that continuing would cause even more costs (as compared to alternatives; Arkes & Blumer, 1985). This is related to the effect of loss aversion (Novemsky & Kahneman, 2005) where losing the initial investment weighs more than the future investments. For example, people tend to have problems replacing old and energy inefficient technology if they just recently have invested, for example, in repairing it. Regret avoidance refers to the effect that people try to avoid the feeling of regret after decisions that turn out to be wrong. Here, research shows that bad consequences of new decisions weigh stronger than the regret of older decisions, which favors again sticking to the status quo (Kahneman & Tversky, 1982). Having purchased a car and then realizing that maybe an electric bike would have been the better decision might lead to retrospectively justifying the car purchase decision. The closely related phenomenon cognitive dissonance (Festinger, 1962) is an aversive feeling that arises if several cognitions a person has at the same time do not fit together. For example, a car owner might consider her-/himself a dedicated environmentalist, which leads to incongruent cognitions around car ownership. This threatens the desire of people to be consistent in their cognitions (Festinger, 1962). Especially for prior decisions which lead to observable outcomes, it is difficult to get out of this aversive state other than justifying the old decision. This then again leads to a status quo bias where previously taken decisions are upheld. Self-perception theory (Bem, 1972), finally, assumes that people often form their attitudes by observing their own behavior in the past, particularly attitudes that are not of central importance for them. The mechanism again builds on the human desire for consistency. Observing one's own decisions in the past to infer personal attitudes to subsequently use these attitudes to make future decisions leads to stability in behavior and resistance to change from status quo.

2.4 Habits and Routines

A factor that has received much attention when analyzing mechanisms behind inaction or the inability to change behavior is habits and behavioral routines (Klöckner & Verplanken, 2012). In short, the basic assumption is that in cases where behavior is successfully repeated in stable situational conditions, humans develop cognitive structures which allow to automatically perform behavior in a given situation without having to dedicate higher-order thinking to it. This saves cognitive resources which then can be dedicated to other tasks without jeopardizing successful performance of a behavior. The performance of the behavior gets associated with situational triggers which shortcuts active thinking (Ouellette & Wood, 1998). This basic mechanism of human decision-making is important for our daily functioning (imagine you would need to think through every morning how to prepare breakfast or brush your teeth), but it also comes at a cost: as long as the automatic performance of the behavior delivers reasonably satisfying results, behavior will not be adjusted (Verplanken & Wood, 2006). Even more important, for behaviors that are strongly habitualized, only a very limited amount of external information is processed (Aarts et al., 1997). In other words, when habits become strong, they make our lives easier on the one hand, but on the other hand, they also make us irresponsive to changes in the environment that might tell us that our behavior is no longer the best choice (for us or for the environment).

There are numerous studies showing that strong habits are an important barrier for changing behavior if they act against pro-environmental intentions that might have formed (Klöckner & Matthies, 2004; Russell et al., 2017; Verplanken et al., 1994). On the other hand, if pro-environmental behavior becomes habitualized, the performance of this behavior does not require a large amount of cognitive effort any longer. Thus, deactivating existing non-environmental habits, actively changing behavior through intentional decisions, and then establishing new habits should be a promising strategy for environmentally relevant behaviors that are performed often and under relatively stable contextual conditions. So far, there is limited evidence of successful attempts to implement this strategy, however.

If strong habits make us immune to detect meaningful changes in decision situations, how can we override them and change our behavior? One promising way of habit change has been studied under the so-called habit discontinuity hypotheses (Verplanken & Roy, 2016). The assumption in this approach is that if habits are strongly linked to situational triggers, then disrupting this link should be a way of deactivating habits and open a time window in which interventions targeting intentions should have a better chance to work. This disruption can be achieved by either changing the context significantly enough so that the behavioral triggers no longer exist, or the successful performance of the behavior is disrupted (Verplanken & Wood, 2006). An example for the first approach is addressing people who are in periods of change (e.g., moving house, starting a family, changing job) assuming that their established habits are naturally not triggered anymore. This line of argument has been followed in a number of studies showing that life events trigger changes in mobility patterns (Klöckner, 2004) or consumption patterns (Schäfer et al., 2012) or that citizens moving into a city are more responsive to interventions in their mobility behavior (Bamberg, 2006). In a study of the behavior change potential in citizens who recently moved to a city (as compared to citizens who lived there for a long time already), Verplanken and Roy (2016) found that indeed the people who were in a changing life situation were more likely to respond with behavioral change to a "pro-environmental goodie-bag" and information intervention than established residents. In a recent paper by Richter et al. (2021) on the effects of COVID-19 on coastal communities in five different biosphere reserves around the world, the authors found that the outbreak of the global pandemic has the potential to serve as a window of opportunity for sustainable behavior change as it also interferes with established habits. On a larger scope, societies or organizational structures can be pushed to establish new laws or initiate peace processes during a window of opportunity after experiencing major disasters as described in Birkmann et al. (2010). They also were able to determine that this window of opportunity for change seems to last about 3 months. After that, habits begin to establish again.

Another line of theories related to automatic versus deliberate decision-making are dual process theories (Evans, 2009). A common assumption in this theory family is that humans have two ways of decision-making: one system is fast, automatic, and unconscious and lacks attention to specific details, whereas the other system is deliberate and voluntary and has high precision but is slow and demands cognitive capacity. The first system has a lot of resemblance with what has been described as habits before (though the two approaches are not identical) and has similar effects on processing of information and resistance to change. An interesting question is how the change between the two ways of processing is achieved or what triggers such a transition. In their influential paper, Louis and Sutton (1991) coined the phrase of "switching cognitive gear" from habitual thinking to active thinking. They conclude that three aspects can contribute to switching cognitive gears, namely, novelty (something unexpected or new is happening), discrepancy (what has been done in the automatic way of thinking does not work out as planned or has negative consequences), and requests for active thinking (somebody tells you to spend some thinking on a matter). The first two are in line with the strategies to break habits outline above. The third one is a self-regulated override of automatic thinking. It has been shown before that this is possible but requires constant attention (Aarts et al., 1997). Over time, old habits of thinking might "sneak" back in and steer behavior back to old patterns. Travel mode choices are a very good example for this type of mechanism. In everyday life, an individual might choose the car for their way to work automatically – hence without thinking – every morning. This then would be a system 1 decision. However, gears can be switched to deliberate decision-making if, for example, (a) the person changes the workplace and needs to establish new travel routines, (b) the workplace reduces the amount of parking spaces and the established routines do not succeed anymore to the same degree, or (c) there is a campaign at the workplace to promote bicycling and the person makes an active effort to participate.

Pinder et al. (2018) formulate an interesting combination of modern habit theory, dual process theories, and goal setting theory (Locke & Latham, 2013) which they refer to as the habit alteration model. The model assumes that situational triggers are filtered through type 1 (unconscious) and type 2 (conscious) attention processes (filter phase), which is an assumption taken from dual process theories. Then, both streams prepare actions when the filtered situational cues trigger memory processes (preparation phase). For the automatic type 1 stream, these memory processes refer to the implicit memory and lead to action impulses, whereas the deliberate string of decision-making refer to explicit memory and are open to self-control and self-monitoring processes. This process ends with forming an intention. Both the action impulses and the intention need to cross a threshold to become relevant for the behavioral response (action phase). The response then feeds back on the interpretation of the next situation. Behavioral interventions therefore need to be tailored to the phase of the process and the stream of decision-making (unconscious type 1 or conscious type 2).

2.5 Social Norms and Values

Whereas this chapter so far focused on individual cognitive mechanisms, which contribute to behavioral inertia, we now shift the focus to psychological processes with roots in the social aspects of human being and the role they might play for locking people into the status quo. In the present section, we analyze the role of social norms and values, before moving to social identities in the next section. Social norms are a psychological phenomenon which describes how other people influence our behavior. There are many facets of social influence and social norms which have been studied in social psychological research, but for this chapter, we will mainly focus on two broad types of social norms: descriptive norms and injunctive norms (Cialdini et al., 1991; Thøgersen, 2006).

Injunctive norms are an assessment of what a person thinks people that are important to them might expect them to do would support or sanction (Thøgersen, 2006). If, for example, a person is considering shifting from a (partially) meat-based to a plant-based diet, they will consider what important others think about plant-based diets. Would their family approve? What would colleagues say if they stopped eating meat in the canteen? What about friends on a Friday evening in town? Social psychological theories like the theory of planned behavior (Ajzen, 1985) conceive of this social influence as one of the important determinants of behavior (in case of the theory of planned behavior, social norms are considered together with the person's attitudes and the amount of control the person perceives having over the behavior). In a recent review, Keizer and Schultz (2018) summarize the literature on social norms' influence on pro-environmental behavior and conclude that social norms tend to have a considerable influence on people's behavior and that this influence is often underestimated – especially if self-reports are involved: people tend to report being less receptacle to social influence than studies show that capture social norms unobtrusively. In an interesting cross-cultural study, Culiberg and Elgaaied-Gambier (2016) show that the translation of national pro-environmental social norms (which might be understood as the cultural level of pro-environmentalism in a country) was translated to individual pro-environmental behavior via the mediating variable of social norms but that the strength of this impact was depending on the country the data was collected in (weaker influence in France than in Slovenia). This shows that the degree to which social norms impact behavior depends to a certain degree on the cultural context.

A second important type of social norms is descriptive norms. Whereas injunctive norms capture the normative message a person extracts from other people's communication ("what should be done"), the descriptive norm captures what a person can actually observe other people are doing ("what is done"). If many people can be observed recycling, the descriptive norm is that one should be recycling and that recycling is the normal and accepted behavior in the observed situation. Vice versa, if many people are observed littering in the park, this behavior becomes the accepted norm (Reno et al., 1993). In a study on how pro-environmental behavior transfers between adolescents and their family members, Grønhøj and Thøgersen (2012) found that the dominating norms in a family strongly influence the adolescents' pro-environmental behavior and that especially the influence of descriptive norms is strong: the adolescents adjust their own behavior to what their parents were doing (and to a lesser degree to what they were telling them to do). This finding was replicated and extended in a study by Wallis and Klöckner (2018), who found that there is a particularly strong agreement between children's and parents' pro-environmental behavior for behaviors that are easier to observe (because they are performed for example in the living room), as compared to behaviors that are less observable (because they happen in the child's room). Again, this underlines the – often underestimated – power of descriptive social norms.

While we have argued that social norms in both forms (descriptive and injunctive) can have a strong influence on people's behavior, this influence is of course not per definition positive of negative from an environmental perspective. If the social norms are in favor of pro-environmental behaviors, they can make pro-environmental behaviors widespread and "force" people to comply with them (Cialdini & Jacobson, 2021). In most cultures, littering in public spaces is not socially accepted; thus there are rather strong social sanctions to fear if that norm is violated, especially when other people can observe this violation. However, if social norms are supporting the less environmentally friendly behavior (as is often the case for meat-heavy diets, individual car use, or frequent air travel to give some examples), social norms can be an important barrier for change. Changing behavior then means going against what other people expect you to do or would approve of, and against what they do themselves, which is an aversive state of mind (Berthoz et al., 2002). Van Kleef et al. (2015) systematically analyzed the mechanisms how violations of norms are processed both by the observers (with anger, derogation, sanctioning, or lowering the social status of the transgressor) and the transgressor (with feelings of guilt or shame). However, since protecting the environment on an abstract level is a widely supported in almost all societies (Kenny, 2021), the social sanctions for norm-violating pro-environmental behavior might be more subtle. Members of the majority which is stabilizing the status quo will likely not attack a person for being environmentally friendly but rather underline that the person's behavior is too extreme, is not the most efficient way to solve the problem, or is not possible for everyone.

An interesting question is how social norms can change if the prevalent norms have such a strong regulating character. One answer is that social norms can be challenged also by minorities and that – over time – their behavior can lead to changes also of the prevalent norms in societies. Wood et al. (1994) conducted a systematic meta-analysis of minority influence experiments and found that minority opinions voiced through an appeal in fact changed the opinions of respondents in the direction of the minority appeal to some degree. This effect was interestingly stronger in situations when the respondents' opinions were measured without the majority being able to observe them and in aspects indirectly related to the attitude object (e.g., additional features of the object people were surveyed about). Both findings indicate that deviation from the majority opinion starts where it is less likely to be detected by representatives of the mainstream opinion. Furthermore, Wood et al. (1994) found the minority influence to be stronger if the minority argued consistently and if the number of minority members increased. These aspects tell us that consistently arguing minorities of a certain size can in fact have an impact which draws the majority in their direction, first in the less socially costly unobservable domain and on side aspects of the behavior and later then moving into the observable domain. In an experiment, Demarque et al. (2015) found that communicating the pro-environmental choices of even a minority of buyers in an online shop is able to move the majority of buyers into a (slightly) more pro-environmental buying behavior.

In the paragraphs above, we have already hinted at that general values nowadays are already in favor of pro-environmental behavior, without that being necessarily visible in people's behavior. In psychology, values are understood as guiding principles in life, which are relatively stable and provide the moral benchmark for more specific behavioral decisions. There are several well-studied taxonomies of value orientations and corresponding studies on distributions of value orientations worldwide (e.g., Inglehart & Welzel, 2005; Schwartz, 2006). In the Schwartz value system, for example, ten basic value orientations are ordered in a circular structure, where neighboring value orientations in the circle have a higher likelihood to be held by the same individuals than value orientations on opposite sides of the circle (Schwartz, 2012). Higher tendencies for pro-environmental behavior have been associated particularly with values from the self-transcendence sector of the circle, namely, universalism and benevolence (Schwartz, 2007). Universalism refers to prioritizing welfare of other people and nature; benevolence means prioritizing welfare of people with whom one has frequent contact. However, even though values have been shown to impact pro-environmental behavior broadly and have been proposed as one mechanism behind spill-over effects from one behavioral domain to another (Thøgersen & Ölander, 2003), the link between values and behavior is only indirect, with many potentially interfering variables (Klöckner, 2013a, b; Stern, 2000). Most importantly, it cannot be assumed that values such as universalism are always triggered in situations where people make decisions with environmental consequences. A critical aspect is, therefore, that a situation is perceived by the deciding person as morally relevant, which then triggers the value orientations a person embraces. On the opposite, social dominance orientation, hence the support for hierarchies and inequality in societies, has been shown to correlate negatively with pro-environmental actions (Milfont et al., 2018) and positively to climate change denial (Jylhä et al., 2016).

2.6 Social and Individual Identities

Closely linked to values and norms is the aspect of identity and its role for our behavioral choices. In psychology, the concept of identity captures the need of individuals to define who they are based on the categories they place themselves in. In identity theory this is referred to as identification; in social identity theory this is described as self-categorization, or which groups we feel to be part of (Stets & Burke, 2000). One could, for example, define oneself by being a worker in the automobile industry, by being a meat eater, by being a passionate car driver, or by being a conservative. One can define oneself by being a member in the union of oil workers, a political party, a protest movement against carbon taxes, or a racing club. On the other hand, one can define oneself as an environmentalist, a vegan, or a bicyclist. One could identify oneself with being a member of an environmental organization, a community gardening project, or Extinction Rebellion. All of these individual and social identities will have an obvious relation to environmental behavior (either positively or negatively; Udall et al., 2020, 2021). However, even seemingly neutral individual and social identities might develop a relevant impact on environmental behavior: if being a fan of a certain rock musician is an important part of my identity and this musician suddenly makes a stand for climate action, this might influence my own decisions. Or if the football team I play in suddenly becomes part of an alliance of football players for the climate, this will certainly also color the way I see the topic (see, e.g., www.weplaygreen.com or www.pledgeball.org).

In fact, several recent reviews and meta-analyses underline the consistent effect of different types of individual and social identities on environmental action (Masson et al., 2017; Udall et al., 2020, 2021). Nigbur et al. (2010) found, for example, that self-identification as a recycler was the second strongest predictor of intentions to recycle after attitudes, underlining that self-identification can be a powerful variable. Overall, individual identities seem to be more strongly linked to pro-environmental behavior than social identities (Masson et al., 2017), but in case the pro-environmental behavior is social, the influence of social identities becomes relatively stronger.

However, identities are to a certain degree fluid. We change the parts of our (especially social) identity that we flag depending on the social context we are in (Stets & Burke, 2000). Whereas our intellectual sophistication might be important for us while we discuss in the book club we are in, this aspect might be less important when we are part of the supporters of our local football club or while we are in the audience of a rock festival. Under these circumstances, other parts of our social identities are important. In social psychology, a person's self is assumed to be a patchwork of different individual and social identity pieces, which can vary in salience and centrality (Stryker & Serpe, 1994). Some aspects of your identity are very central for who you are and are important and salient across many situations, while others are not. Among these noncentral identity pieces, salience might be increased depending on context influences.

What makes identity an interesting concept to look at in the context of this book is not only that salience of identity components may be changed but also that people react strongly to threats toward their identities, especially the central components. In a study on changing travel behavior, Murtagh et al. (2012) found that the threat to people's self-identity as motorist was a factor that explained resistance to change over and above processes like habits or reactance. In another study, links were made between talking about air pollution in a neighborhood and the experienced threat of the residents' place identity (Hugh-Jones & Madill, 2009). Respondents in this study felt that their place identity was threatened by the study about local air pollution they were part of, leading them to defend their decision to live there, consequently downplaying the pollution levels. Thus, threatening central parts of people's identities might cause resistance and cognitive mechanisms to redefine the identity threatening input.

2.7 Lock-in Situations

Before we analyze how disruptive communication may be helpful to overcome the psychological barriers for change outlined in the sections above, we will present a phenomenon on a higher level that often leads to the inability of a social system to change, namely, "lock-in." This phenomenon describes a situation where two or more products (or behaviors) with initially equal costs and benefits over time develop markets where one of the products either locally or globally dominates (Liebowitz & Margolis, 1995). Examples for this effect are manifold, as Blu-ray dominating the high-resolution DVD market after some initial battle with alternative formats or battery-electric cars strongly dominating the market for electric cars at the moment at the expense of hydrogen driven cars. Often rather random events lead to one of the technologies getting the initial slight kick start. Janssen and Jager (1999) could show with an agent-based simulation of such lock-in situations that a number of psychological processes can account for the effects. They found that if there is a strong need for connecting the product to one's identity, the comparison with other people's behavior becomes initially very important, until the choices stabilize and individual repetition of previous choices becomes more and more dominant, freezing the behavioral patterns permanently. In the early days of electric cars in the Norwegian market, the influence of decisions of people close to you might thus have been very important, whereas in the now more mature market structure, earlier choices for battery-electric vehicles are just repeated, thus leading to a lock-in. This combination quickly leads to lock-in situations for products with small differences. If, on the other hand, the need for satisfying personal taste is more important, decisions are made more deliberate, and lock-in is less likely to develop (Janssen & Jager, 1999).

A different understanding of a lock-in is proposed by Ivanova et al. (2018) and Maréchal (2010). In both papers, behavioral lock-in is conceptualized by limited agency of the deciding individuals as the decisions are to a large degree determined by the available (and not available) infrastructure. Examples are high-energy consumption in the building sector caused by a certain configuration of building techniques and settlement structures, or high-energy mobility patterns being caused by decentralized settlements and limited access to alternative travel modes. Although these structural impacts on individual decisions are nonpsychological in the first place, they are reflected in the psychology of decision-making by discounting or excluding certain behavioral alternatives already early in the process.

What both types of lock-in have in common is that it is very demanding to overcome them as the number of available alternative behaviors at reasonable costs is very restricted once a lock-in situation occurred.

2.8 Disruptive Communication in Relation to the Psychological Mechanisms of Inaction

In the previous sections, we have described a number of psychological mechanisms that contribute to behavioral inertia and resistance to change. In this section, we will now explore how disruptive communication as defined in Chap. 1 of this book might be helpful in overcoming these barriers but also if some of these mechanisms might hamper the success of disruptive communication. Since the concept of disruptive communication as defined in our book has not been tested empirically, the

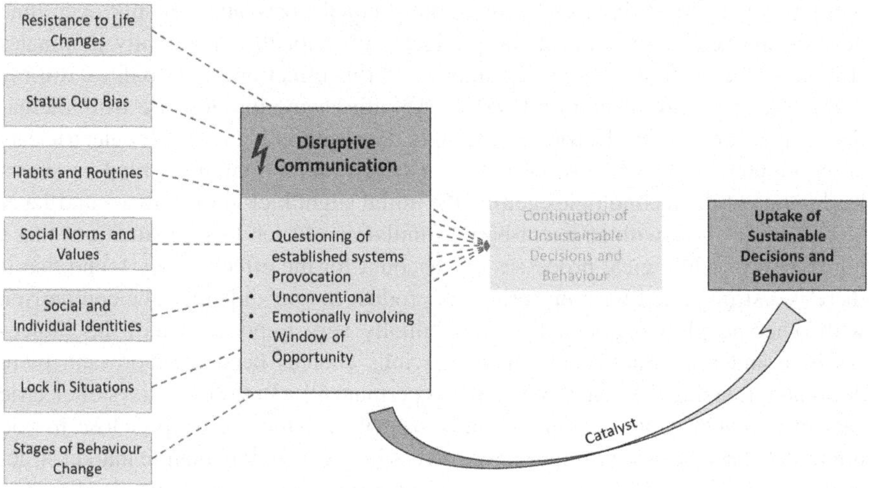

Fig. 2.2 Potential pathway of how disruptive communication can overcome behavioral inertia

following paragraphs should be treated as hypotheses. Figure 2.2 gives an overview about the different mechanisms by which we expect disruptive communication to work.

2.8.1 Change as Stress

In Sect. 2.2, it was outlined that substantial changes of life situations often cause stress responses in people. As disruption is a stressor, *disruptive communication might elevate stress levels* of people, depending on how disruptive the communication element will be experienced. When designing the disruptive communication measure, it is thus important to design it in a way that gives people an outlet for potential stress reactions.

With respect to the concept of disruptive communication, these considerations indicate that there is a thin line to be aware of here. Causing disruption may have positive effects, if it triggers peoples' interest to change and stops routine live (see also Sect. 2.4). Causing chronic worry or distress will have negative physiological and mental effects, which need to be avoided in a disruptive communication strategy. However, the nature of disruptive communication as outlined in Chap. 1 is usually nonpermanent; therefore negative chronic effects are unlikely. Following Cohen and Hoberman (1983), designing successful disruptive communication can be achieved by making sure that the disruptive communication strategy also includes a component of positive experiences (e.g., hope or awe) and that it builds on or creates social support mechanisms in the stage following the disruption. Many of the examples of disruptive communication presented in the remaining chapters of this book have this quality: on the one hand, they disrupt people's lives and are potential sources for stress; they have qualities that upset people. On the other hand, they also have elements of art, playfulness, social components, or other features that balance these stress triggering components and act as catalyst for positive change. However, there are interpersonal differences in how much stress is adequate stress (Tonhajzerova & Mestanik, 2017). Whereas one person might appreciate the artistic quality of the disruptive element, another might be just provoked and very angry. Therefore, the second element of disruptive communication outlined in Chap. 1, namely, following people up after the disruption, becomes very important. Stress management (e.g., Greenberg, 2012) needs also to be part of this follow-up.

2.8.2 Status Quo Bias

As argued in Sect. 2.3, humans have a tendency to favor the status quo over change for many reasons. The following paragraphs analyze how disruptive communication might be helpful in overcoming the status quo bias. Since the first mechanism behind the status quo bias is the rationality of avoiding transition costs, the first benefit of disruptive communication might be *to make the costs of none-transition more salient*. If the communication event highlights (and maybe contrasts) costs of none-transition vs. transition in a creative way, this might reduce the tendency to rationalize none-action. Staging or co-creating potential future scenarios in a low-carbon world in a tangible way might also reduce the uncertainty connected to the transition, which again reduces status quo bias fuelled by this mechanism. Sheppard (2012) describes some strategies for how to achieve this in his book about visualizing climate change.

An interesting question is if disruptive communications can be designed that address the status quo effect caused by decisional heuristics. In a recently published paper, Gal (2021) argues that the endowment effect does not reflect that people value losses more than gains but rather that people favor inaction over action, because they fear that actively doing something might cause failures. Omitting to act is experienced as a safer option, because there is no clear agency of the individual. Here, disruptive communication might come in, making the agency of none-action tangible. Thus, *disruptive communication might be helpful to visualize that inaction is also an active choice*. For example, an installation visualizing or making the outcome of none-action in the climate domain tangible and contrasting that with future scenarios where action has been implemented could be helpful. Or an approach that shows that inaction actually means the choice of status quo alternatives: not changing the heating system means choosing the old heating system.

With regard to reducing the effects of the anchoring heuristic described above, research has shown that providing people with information about the existence of the anchoring effect can be helpful (Shepperd et al., 2018). However, other research shows that this is difficult to transfer from a laboratory context to real life (Fuesting, 2014), as the expected effects of information about the anchoring mechanism did not change people's receptiveness for it. Interestingly, the author of the latter study explicitly mentions that the field intervention to reduce the anchoring effect was designed offering only little disruption to an individual's daily routine. Maybe, this aim not to disrupt people's lives can be an explanation for the lack of effect. What if the information was designed as a disruptive communication which also provides people with an alternative anchor to relate to? What if the provided anchor was the disruptive element to shock people and makes them stop? Our hypothesis is that *disruptive communication could be a way to address status quo effects caused by anchoring by actively holding up the anchor "into people's faces."*

As also cognitive dissonance was named as one possible cause of a status quo bias, it is relevant to analyze how disruptive communication might disturb the balance of cognitions and thus both induce and resolve cognitive dissonance. According to dissonance theory (Festinger, 1962), people experience an aversive feeling if there is a mismatch between two thoughts they have at the same time (e.g., "I am an environmentalist" and "I like flying to Paris for a weekend trip") or between their thoughts and values and their behavior. In this context, *disruptive communication can clearly be a cause for cognitive dissonance* by pointing out this mismatch (e.g., by highlighting the mismatch between people's stated values and behavior). However, *disruptive communication might also resolve a situation of cognitive dissonance* by adding weight to one side of the imbalance. If, for example, a person is torn between the wish to be an environmentalist and to go on a weekend trip to Paris, the scales might tip to the environmentalist side when they are disrupted by a communication that highlights environmental issues. An example could be the booking page of the travel agency of our university, which displays the message "Often trips inland are possible by train, have you checked the connections?" together with a link to the section of the webpage, where you can book the train, when you are searching for inland flight connections.

2.8.3 Habits and Routines

Habits and routines are probably the psychological mechanism behind resistance to behavior change which is easiest to link to disruptive communication. The very essence of habits is that they link behavioral response patterns to situational triggers and take behavior out of reach for deliberate decision-making, as was outlined in Sect. 2.4. What deactivates habits are disruptions of the automatic performance. Thus, *disruptive communication might be an excellent tool to deactivate habits* at least temporarily, opening a window of opportunity for change. For this to work effectively, disruptive communication would either need to stop the performance of the behavior (like, e.g., a roadblock by Extinction Rebellion would do), or it would need to trigger the attention of the person enough to motivate them to "switch" their cognitive gears to active thinking. To deactivate habits, something needs to happen that is a strong enough stimulus to break through the reduced information processing and that motivates people enough to spend cognitive energy to make deliberate decisions (Verplanken & Wood, 2006). Both are features of disruptive communication as outlined in Chap. 1. Currently, researchers discuss the role of the COVID-19 pandemic as a potential system disrupter providing a time window for sustainable change making besides its obvious disastrous consequences (Richter et al., 2021; Schmidt et al., 2021). Artificially creating a disruption that is strong enough to change habitual behavior but does not come with long-lasting negative side effects is a fine line disruptive communication research sets about to explore.

2.8.4 Social Norms and Values

Disruptive communication like any form of communication is made by someone, which makes it a form of social influence. However, since it is disruptive, it is likely not the majority position or does not represent the majority attitude or behavior. As such, it is an example of a potential minority influence on the formation of the majority norms. Disruptive communication does that in a way that "sticks out"; it violates the implicitly shared group rules on how communication should be. At the first glance, this might reduce rather than increase its impact (see also Chap. 4), as people might perceive it as inappropriate. However, if implemented in a creative way, it may also feel empowering and become a catalyst for members of a silent minority (or even majority). By showing that a very outspoken person (or persons) has a similar position, these silent people might be motivated to become more vocal. A very good example for this is the climate strike initiated by Greta Thunberg. Implementing an element of disruptive communication (the climate strike in front of the Swedish parliament), she encouraged many young people to stand their ground against the lack of climate measures in their countries. This made demanding more engaged climate policies not only an accepted but an appreciated norm for the young generation and beyond. Thus, *disruptive communication has the potential to challenge majority norms, to make minority positions tangible and visible, and to become crystalizing points for people being in a silent opposition to majority norms.*

With respect to values, we have argued earlier in this chapter that many people already hold pro-environmental values in an abstract form. However, these values are seldom acted upon, because they are not salient in situations when people make decisions. If disruptive communication happens at points in time where people make decisions about their behavior, they might make the link to the values people hold visible and put the finger on the mismatch between people's value orientations and their behavior. This situation might then cause a feeling of dissonance which can be solved by adjusting the behavior (in the best case). Thus, *disruptive communication might also be a tool to link people's behavior closer to their values* in everyday situations.

2.8.5 Social and Individual Identities

When analyzing the relationship between disruptive communication and identities, the effect of the communication action can go in several directions. In case the disruptive communication is in line with one's personal or social identity, *disruptive communication will strengthen the identity and make it salient* in a given situation. So, if the communication, for example, addresses the food waste problem in a canteen, you are in the canteen at the buffet, and you already identify yourself as a person for whom it is important not to waste anything (especially food), your identity becomes salient through the disruptive trigger. If you identify yourself as being a citizen of your hometown and the disruptive event makes the link between being a citizen of that town and reducing the impact of your daily travels, your identification with the town might make you responsive for the message of the disruptive event. *Creative disruptive events might also be a means to form social identities*: if we take Extinction Rebellion as an example, many of their actions include an artistic reflection and expression, which then again binds the group together as a common identity. In a study, Hess (2019) found that singing and song writing had a strong identity forming function for activists, for example.

However, *identities can also be threatened by a disruptive event*: imagine you are a proud driver of a vintage car and driving on pleasure trips with that car on a warm summer day is what is an important indicator of who you are, then a disruptive event like a road blockage questioning car driving in general and pleasure trips in particular might be experienced as a threat to your identity. If not addressed, this identity threat might become a powerful source of reactance and anger. Therefore, designing disruptive communication should focus on managing the identity threatening potential by, for example, being compatible with several identities or, as stated before, provide an outlet for negative emotions and the opportunity to channel them into constructive solution development.

2.8.6 Lock-in

The final aspect discussed in the chapter was lock-in, thus the loss of agency by the individual and the domination of a market situation or political-structural situation. Disruptive communication will not break lock-in situations right away; however, it might be a way to draw attention to lock-ins and might motivate people to change structures to break up lock-ins.

2.9 Different Stages of Behavioral Change

Before concluding this chapter, it makes also sense to take a look at how people change their behavior. Behavior change is not a sudden shift from one behavior to the other. Behavior change occurs over time and is characterized by people moving between stages (Gollwitzer, 1990; Schwarzer, 2008). To model and explain people's shift toward environmentally relevant behaviors, Bamberg developed the stage model of self-regulated behavioral change (Bamberg, 2013). This model combines the model of action stages (Gollwitzer, 1990) and static, driver-oriented models used in environmental psychology. The model of self-regulated behavioral change comprises four action stages (predecisional, preactional, actional, and postactional) which are understood as stages of decision-making. People move in between these stages by developing specific intentions: goal intentions, behavioral intentions, and implementation intentions. To transition from the first stage, predecision, to the second stage, preaction, the current behavior needs to be perceived as problematic, leading to an intention to change this behavior (goal intention). Subsequently, an alternative behavior is identified, and the intention to perform this newly chosen behavior is created (behavioral intention). Next, this new behavior is repeatedly performed depending on how strong a person's implementation intention is, which can be supported by good planning abilities and self-efficacy. The postaction stage represents the phase in which the new behavior is either fully established or the person relapses back into a previous stage or behavior. In a recent review by Keller et al. (2019), the authors concluded that the general framework of the stage model of self-regulated behavioral change is supported empirically across various behavioral domains. Interventions with stage-tailored information are more effective in fostering stage progression and behavior change as compared to non-tailored information, across a variety of behavioral domains such as mobility and food consumption.

For behavior change, this model implies that each transition from one stage to the other requires stage-specific interventions, matching the current communicative needs of the target group. Practically, this means that effective interventions require knowledge about the behavioral stage in which the target group is currently in. This comes with a variety of practical shortcomings such as reliance on accurate self-report of the respondents as well as inter-individual differences. In their review, Keller et al. (2019) discuss these shortcomings of the model in terms of its practical manageability and comparability.

Disruptive communication can be a way to approach the model's practical limitations by providing a technique that directly transfers people into the actional stage, independent of their jump-off point. By using disruptive communication, stage assessment and tailored interventions for each stage and target group might become obsolete: instead of gently nudging people from one stage of behavioral change to the next, disruptive communication is designed to create a strong enough impulse to urge action directly.

As an example, the original model suggests that behavioral interventions to motivate the reduction of beef consumption should be adapted to the consumer's stage of behavioral change (Klöckner, 2017). Consumers in the predecision stage might benefit most from information about environmental problems caused by meat consumption. In the decisional stage, appropriate behavioral alternatives need to be lined out. In the action stage, specific implementation advice is necessary to provoke the behavior to be carried out. In the postaction stage, support in avoiding relapse into previous behavioral patterns is needed. With disruptive communication, it might be possible to provoke action independent of the consumer's stage in the model, provided a behavioral alternative is immediately provided. Practically, graphic pictures showing negative consequences of meat consumption (e.g., animals being slaughtered or burning rainforest) could be printed on meat products, similar to the images printed on cigarette packets. Presumably, this might cause a shock to consumers. Placing a comparable but more sustainable alternative directly next to the meat (such as a plant-based meat replacement) provides an immediate solution. We claim that due to its strong emotional components, this technique will be effective for consumers across all stages of behavioral change.

Similar to previously discussed techniques in this chapter, the success of this method depends on the careful and immediate channelling of the emotions evoked during the disruption into behavioral solutions.

2.10 Conclusion

In this chapter, we have analyzed a number of individual, social, and structural barriers against transformative change of behaviors. This list is, of course, far from being comprehensive and dictated by our own research interests. However, based on the analysis of psychological mechanisms that lead to this behavioral inertia, we also argued for how the concept of disruptive communication we outlined in Chap. 1 might be an alternative, sometimes more promising way of addressing these barriers. Since the concept is new, most of the assumptions and hypotheses formulated in the previous section are not empirically tested, yet. It is, for example, unclear where the line goes between activating people and motivating them to go into a discussion and negotiation process on the one hand and irritating them so much that they avoid further engagement. Keeping this in mind, we will in the following chapters study a number of applications of something that could be referred to as creative disruptive communication from different angles to underline our hypotheses formulated in this chapter.

References

Aarts, H., Verplanken, B., & Van Knippenberg, A. (1997). Habit and information use in travel mode choices. *Acta Psychologica, 96*(1), 1–14.

Ajzen, I. (1985). From intentions to actions: A theory of planned behaviour. In *Action control* (pp. 11–39). Springer.

Arkes, H. R., & Blumer, C. (1985). The psychology of sunk cost. *Organizational Behaviour and Human Decision Processes, 35*(1), 124–140.

Bamberg, S. (2006). Is a residential relocation a good opportunity to change people's travel behaviour ? Results from a theory-driven intervention study. *Environment and Behaviour, 38*(6), 820–840.

Bamberg, S. (2013). Changing environmentally harmful behaviours: A stage model of self-regulated behavioural change. *Journal of Environmental Psychology, 34*, 151–159.

Bem, D. J. (1972). Self-perception theory. In *Advances in experimental social psychology* (Vol. 6, pp. 1–62). Elsevier.

Berthoz, S., Armony, J., Blair, R., & Dolan, R. J. (2002). An fMRI study of intentional and unintentional (embarrassing) violations of social norms. *Brain, 125*(8), 1696–1708.

Birkmann, J., Buckle, P., Jaeger, J., Pelling, M., Setiadi, N., Garschagen, M., et al. (2010). Extreme events and disasters: A window of opportunity for change? Analysis of organizational, institutional and political changes, formal and informal responses after mega-disasters. *Natural Hazards, 55*(3), 637–655.

Cialdini, R. B., & Jacobson, R. P. (2021). Influences of social norms on climate change-related behaviours. *Current Opinion in Behavioural Sciences, 42*, 1–8.

Cialdini, R. B., Kallgren, C. A., & Reno, R. R. (1991). A focus theory of normative conduct: A theoretical refinement and reevaluation of the role of norms in human behaviour. In *Advances in experimental social psychology* (Vol. 24, pp. 201–234). Elsevier.

Clayton, S. (2020). Climate anxiety: Psychological responses to climate change. *Journal of Anxiety Disorders, 74*, 102263.

Cohen, S., & Hoberman, H. M. (1983). Positive events and social supports as buffers of life change stress 1. *Journal of Applied Social Psychology, 13*(2), 99–125.

Culiberg, B., & Elgaaied-Gambier, L. (2016). Going green to fit in–understanding the impact of social norms on pro-environmental behaviour, a cross-cultural approach. *International Journal of Consumer Studies, 40*(2), 179–185.

Demarque, C., Charalambides, L., Hilton, D. J., & Waroquier, L. (2015). Nudging sustainable consumption: The use of descriptive norms to promote a minority behaviour in a realistic online shopping environment. *Journal of Environmental Psychology, 43*, 166–174.

Evans, J. S. B. (2009). How many dual-process theories do we need? One, two, or many? In J. S. B. Evans & K. Frankish (Eds.), *In two minds: Dual processes and beyond* (pp. 33–54). Oxford University Press.

Festinger, L. (1962). *A theory of cognitive dissonance* (Vol. 2). Stanford University Press.

Fuesting, M. A. (2014). Come on down: Investigating an informational strategy to debias the anchoring heuristic. *CrissCross, 2*(1), 5.

Gal, D. (2021). Why the sun will not set on the endowment effect: The endowment effect after loss aversion. *Current Opinion in Psychology, 39*, 12–15.

Gollwitzer, P. M. (1990). Action phases and mind-sets. *Handbook of Motivation and Cognition: Foundations of Social Behaviour, 2*, 53–92.

Greenberg, J. (2012). Comprehensive stress management. McGraw-Hill Education.

Grønhøj, A., & Thøgersen, J. (2012). Action speaks louder than words: The effect of personal attitudes and family norms on adolescents' pro-environmental behaviour. *Journal of Economic Psychology, 33*(1), 292–302.

Hess, J. (2019). Singing our own song: Navigating identity politics through activism in music. *Research Studies in Music Education, 41*(1), 61–80.

Hickman, C., Marks, E., Pihkala, P., Clayton, S., Lewandowski, R. E., Mayall, E. E., et al. (2021). Climate anxiety in children and young people and their beliefs about government responses to climate change: A global survey. *The Lancet Planetary Health, 5*(12), e863–e873.

Hugh-Jones, S., & Madill, A. (2009). The air's got to be far cleaner here: A discursive analysis of place-identity threat. *British Journal of Social Psychology, 48*(4), 601–624.

Inglehart, R., & Welzel, C. (2005). *Modernization, cultural change, and democracy: The human development sequence*. Cambridge University Press.

Irwin, M., Patterson, T., Smith, T. L., Caldwell, C., Brown, S. A., Gillin, J. C., & Grant, I. (1990). Reduction of immune function in life stress and depression. *Biological Psychiatry, 27*(1), 22–30.

Ivanova, D., Vita, G., Wood, R., Lausselet, C., Dumitru, A., Krause, K., et al. (2018). Carbon mitigation in domains of high consumer lock-in. *Global Environmental Change, 52*, 117–130.

Janssen, M., & Jager, W. (1999). An integrated approach to simulating behavioural processes: A case study of the lock-in of consumption patterns. *Journal of Artificial Societies and Social Simulation, 2*(2), 21–35.

Jylhä, K. M., Cantal, C., Akrami, N., & Milfont, T. L. (2016). Denial of anthropogenic climate change: Social dominance orientation helps explain the conservative male effect in Brazil and Sweden. *Personality and Individual Differences, 98*, 184–187.

Kahneman, D., & Tversky, A. (1982). The psychology of preferences. *Scientific American, 246*(1), 160–173.

Kahneman, D., Knetsch, J. L., & Thaler, R. H. (1991). Anomalies: The endowment effect, loss aversion, and status quo bias. *Journal of Economic Perspectives, 5*(1), 193–206.

Keizer, K., & Schultz, P. W. (2018). Social norms and pro-environmental behaviour. In *Environmental Psychology: An Introduction* (pp. 179–188). Groningen University.

Keller, A., Eisen, C., & Hanss, D. (2019). Lessons learned from applications of the stage model of self-regulated behavioural change: A review. *Frontiers in Psychology, 10*, 1091.

Kenny, J. (2021). *The evolution of environmental concern in Europe*. Edward Elgar Publishing.

Klöckner, C. A. (2004). *How single events change travel mode choice–a life span perspective.* Paper presented at the 3rd international conference on traffic and transport psychology, Nottingham, UK.

Klöckner, C. A. (2013a). A comprehensive model of the psychology of environmental behaviour – A meta-analysis. *Global Environmental Change, 23*(5), 1028–1038.

Klöckner, C. A. (2013b). How powerful are moral motivations in environmental protection? In K. Heinrichs, F. Oser, & T. Lovat (Eds.), *Handbook of moral motivation* (pp. 447–472). Sense Publishers.

Klöckner, C. A. (2017). A stage model as an analysis framework for studying voluntary change in food choices – The case of beef consumption reduction in Norway. *Appetite, 108*, 434–449.

Klöckner, C. A., & Matthies, E. (2004). How habits interfere with norm-directed behaviour: A normative decision-making model for travel mode choice. *Journal of Environmental Psychology, 24*(3), 319–327.

Klöckner, C. A., & Verplanken, B. (2012). Yesterday's habits preventing change for tomorrow? About the influence of automaticity on environmental behaviour In L. Steg, A. E. van den Berg, & J. I. M. De Groot (Eds.), *Environmental psychology: An introduction* (pp. 197–209): Wiley-Blackwell.

Lehrer, S. (1980). Life change and gastric cancer. *Psychosomatic Medicine, 42*(5), 499–502.

Liebowitz, S. J., & Margolis, S. E. (1995). Path dependence, lock-in, and history. *Journal of Law, Economics and Organization, 11*(1), 205–226.

Locke, E. A., & Latham, G. P. (2013). Goal setting theory: The current state. In E. A. Locke & G. P. Latham (Eds.), *New developments in goal setting and task performance* (pp. 623–630). Routledge/Taylor & Francis Group.

Louis, M. R., & Sutton, R. I. (1991). Switching cognitive gears: From habits of mind to active thinking. *Human Relations, 44*(1), 55–76.

Mallinckrodt, B., Leong, F. T. L., & Kralj, M. M. (1989). Life-change stress and stress symptoms. *Journal of College Student Development, 30*, 332–338.

Marczak, M., Winkowska, M., Chaton-Østlie, K., & Klöckner, C. A. (2021). "It's like getting a diagnosis of terminal cancer." An Exploratory Study of the Emotional Landscape of Climate Change Concern in Norway.

Maréchal, K. (2010). Not irrational but habitual: The importance of "behavioural lock-in" in energy consumption. *Ecological Economics, 69*(5), 1104–1114.

Masson, T., Panno, A., Tiberio, L., Vesely, S., Carrus, G., Fritsche, I., & Klöckner, C. A. (2017). *Identity processes and individual factors in energy decisions – Two comprehensive meta-analyses*. Retrieved from Trondheim: https://www.echoes-project.eu/sites/echoes.drupal.pulsartecnalia.com/files/D4.1.pdf

Milfont, T. L., Bain, P. G., Kashima, Y., Corral-Verdugo, V., Pasquali, C., Johansson, L.-O., et al. (2018). On the relation between social dominance orientation and environmentalism: A 25-nation study. *Social Psychological and Personality Science, 9*(7), 802–814.

Murtagh, N., Gatersleben, B., & Uzzell, D. (2012). Self-identity threat and resistance to change: Evidence from regular travel behaviour. *Journal of Environmental Psychology, 32*(4), 318–326.

Nigbur, D., Lyons, E., & Uzzell, D. (2010). Attitudes, norms, identity and environmental behaviour: Using an expanded theory of planned behaviour to predict participation in a kerbside recycling programme. *British Journal of Social Psychology, 49*(2), 259–284.

Novemsky, N., & Kahneman, D. (2005). The boundaries of loss aversion. *Journal of Marketing Research, 42*(2), 119–128.

Ogunbode, C. A., Pallesen, S., Böhm, G., Doran, R., Bhullar, N., Aquino, S., et al. (2021). Negative emotions about climate change are related to insomnia symptoms and mental health: Cross-sectional evidence from 25 countries. *Current Psychology*, 1–10. https://doi.org/10.1007/s12144-021-01385-4

Ojala, M. (2012). Regulating worry, promoting hope: How do children, adolescents, and young adults cope with climate change? *International Journal of Environmental and Science Education, 7*(4), 537–561.

Ouellette, J. A., & Wood, W. (1998). Habit and intention in everyday life: The multiple processes by which past behaviour predicts future behaviour. *Psychological Bulletin, 124*(1), 54–74.

Panu, P. (2020). Anxiety and the ecological crisis: An analysis of eco-anxiety and climate anxiety. *Sustainability, 12*(19), 7836.

Pihkala, P. (2020). Eco-anxiety and environmental education. *Sustainability, 12*(23), 10149.

Pihkala, P. (2022). Eco-anxiety and pastoral care: Theoretical considerations and practical suggestions. *Religions, 13*(3), 192.

Pinder, C., Vermeulen, J., Cowan, B. R., & Beale, R. (2018). Digital behaviour change interventions to break and form habits. *ACM Transactions on Computer-Human Interaction (TOCHI), 25*(3), 1–66.

Rahe, R. H. (1975). Epidemiological studies of life change and illness. *The International Journal of Psychiatry in Medicine, 6*(1–2), 133–146.

Reno, R. R., Cialdini, R. B., & Kallgren, C. A. (1993). The transsituational influence of social norms. *Journal of Personality and Social Psychology, 64*(1), 104.

Richter, I., Avillanosa, A., Cheung, V., Goh, H. C., Johari, S., Kay, S., et al. (2021). Looking through the COVID-19 window of opportunity: Future scenarios arising from the COVID-19 pandemic across five case study sites. *Frontiers in Psychology, 12*, 2707.

Russell, S. V., Young, C. W., Unsworth, K. L., & Robinson, C. (2017). Bringing habits and emotions into food waste behaviour. *Resources, Conservation and Recycling, 125*, 107–114.

Samuelson, W., & Zeckhauser, R. (1988). Status quo bias in decision making. *Journal of Risk and Uncertainty, 1*(1), 7–59.

Schäfer, M., Jaeger-Erben, M., & Bamberg, S. (2012). Life events as windows of opportunity for changing towards sustainable consumption patterns?. *Journal of Consumer Policy, 35*(1), 65–84.

Schmidt, K., Sieverding, T., Wallis, H., & Matthies, E. (2021). COVID-19 – A window of opportunity for the transition toward sustainable mobility? *Transportation Research Interdisciplinary Perspectives, 10*, 100374.

Schwartz, S. H. (2006). A theory of cultural value orientations: Explication and applications. *Comparative Sociology, 5*(2–3), 137–182.

Schwartz, S. H. (2007). Value orientations: Measurement, antecedents and consequences across nations. In R. Jowell, C. Roberts, R. Fitzgerald, & G. Eva (Eds.), *Measuring attitudes cross-nationally: Lessons from the European Social Survey* (pp. 169–204). Sage.

Schwartz, S. H. (2012). An overview of the Schwartz theory of basic values. *Online Readings in Psychology and Culture, 2*(1), 2307–0919.1116.

Schwarzer, R. (2008). Modeling health behaviour change: How to predict and modify the adoption and maintenance of health behaviours. *Applied Psychology: An International Review, 57*(1), 1–29.

Sheppard, S. R. J. (2012). *Visualizing Climate Change*. Routledge.

Shepperd, M., Mair, C., & Jørgensen, M. (2018). *An experimental evaluation of a de-biasing intervention for professional software developers*. Paper presented at the proceedings of the 33rd annual ACM symposium on applied computing.

Stern, P. C. (2000). New environmental theories: Toward a coherent theory of environmentally significant behaviour. *Journal of Social Issues, 56*(3), 407–424.

Stets, J. E., & Burke, P. J. (2000). Identity theory and social identity theory. *Social Psychology Quarterly, 63*, 224–237.

Stryker, S., & Serpe, R. T. (1994). Identity salience and psychological centrality: Equivalent, overlapping, or complementary concepts? *Social Psychology Quarterly, 57*, 16–35.

Thøgersen, J. (2006). Norms for environmentally responsible behaviour: An extended taxonomy. *Journal of Environmental Psychology, 26*(4), 247–261.

Thøgersen, J., & Ölander, F. (2003). Spillover of environment-friendly consumer behaviour. *Journal of Environmental Psychology, 23*(3), 225–236.

Tonhajzerova, I., & Mestanik, M. (2017). New perspectives in the model of stress response. *Physiological Research, 66*, S173.

Tversky, A., & Kahneman, D. (1974). Judgment under uncertainty: Heuristics and biases. *Science, 185*(4157), 1124–1131.

Tversky, A., & Kahneman, D. (1991). Loss aversion in riskless choice: A reference-dependent model. *The Quarterly Journal of Economics, 106*(4), 1039–1061.

Udall, A. M., de Groot, J. I., de Jong, S. B., & Shankar, A. (2020). How do I see myself? A systematic review of identities in pro-environmental behaviour research. *Journal of Consumer Behaviour, 19*(2), 108–141.

Udall, A. M., de Groot, J. I., De Jong, S. B., & Shankar, A. (2021). How I see me—A meta-analysis investigating the association between identities and pro-environmental behaviour. *Frontiers in Psychology, 12*, 642.

Van Kleef, G. A., Wanders, F., Stamkou, E., & Homan, A. C. (2015). The social dynamics of breaking the rules: Antecedents and consequences of norm-violating behaviour. *Current Opinion in Psychology, 6*, 25–31.

Verplanken, B., & Roy, D. (2016). Empowering interventions to promote sustainable lifestyles: Testing the habit discontinuity hypothesis in a field experiment. *Journal of Environmental Psychology, 45*, 127–134.

Verplanken, B., & Wood, W. (2006). Interventions to break and create consumer habits. *Journal of Public Policy & Marketing, 25*(1), 90–103.

Verplanken, B., Aarts, H., Knippenberg, A., & Knippenberg, C. (1994). Attitude versus general habit: Antecedents of travel mode Choice1. *Journal of Applied Social Psychology, 24*(4), 285–300.

Wallis, H., & Klöckner, C. A. (2018). The transmission of energy-saving behaviours in the family: A multilevel approach to the assessment of aggregated and single energy-saving actions of parents and adolescents. *Environment and Behaviour*. https://doi.org/10.1177/0013916518802342

Wood, W., Lundgren, S., Ouellette, J. A., Busceme, S., & Blackstone, T. (1994). Minority influence: A meta-analytic review of social influence processes. *Psychological Bulletin, 115*(3), 323.

Chapter 3
Activist Art as a Motor of Change? How Emotions Fuel Change

Laura K. Sommer, Erica Löfström, and Christian A. Klöckner

3.1 Art as Catalyst of Change

Art is a powerful medium that has affected all of us, in one way or the other. When asking people about their experiences with art, most of them would be able to tell you about one specific artwork that mesmerized them, caught their attention, and "disrupted" them in their daily endeavors. When thinking back, this art experience might even today change their mood, have an influence on their thoughts, and might well have impacted some of their decisions since then. It is therefore not surprising that we assign art a big cultural value in our society and that it is frequently described as catalyst of change (Belfiore & Bennett, 2007).

From the artists' perspective, many of them feel the need to address societal challenges in their work. This is also true for climate change and other environmental problems humankind is facing (Demos, 2016). Environmental art can be traced back to eighteenth-century Romanticism (Martin, 2015), but more contemporary environmental art has its roots in the "Land art" movement of the 1960s (Wolfe & Russell, 2010). Other movements in the same genre include "Earthworks" (Boettger, 2002, p. 9), "Ecological Art/Eco-Art" (Wallen, 2012; Martin, 2015), "visual activism" or "rebel creativity" (Demos, 2016), and "climate change art" (Nurmis, 2016). Especially "climate change art" has experienced a boost and matured in the last 20 years, because of the increasing urgency of the matter (Nurmis, 2016).

The diversity of artworks makes it difficult to find a clear definition of what can and cannot be understood as environmental art. Marks et al. (2017) were among the first to investigate the effect of environmental art on its audience and this chapter follows their definition:

> Environmental art is a diverse field of creative practice that includes works that aim to engender an appreciation of nature, as well as works that 'employ nature as a medium'. It primarily involves art created with the intention of drawing attention to nature and raising awareness of environmental issues. Environmental art can be permanent or ephemeral and

C. A. Klöckner, E. Löfström, *Disruptive Environmental Communication*,
Psychology and Our Planet, https://doi.org/10.1007/978-3-031-17165-9_3

created in any format such as sculpture, photography, multimedia or performance. (Marks et al., 2017, page 1308)

This definition highlights the intention of the artist to raise awareness for environmental issues. However, the degree of political intention between artists and even between artworks of the same artist can vary. It is therefore helpful to place every art piece on a spectrum of "art-like art" vs. "life-like art" (Wallen, 2012), which respectively describes "art created for its own sake" and "art with a political agenda." In this chapter and the Climart project, the focus is on "life-like art," so specifically art that tries to contribute to change.

Another spectrum that the artwork can vary on is the "level of disruptiveness." This chapter focuses on the degree to which the artwork interferes with someone's personal space, time, and choices (see Chap. 1). Often art is exhibited in a museum setting, where the spectator has to make the decision to visit an exhibition on, for example, climate change art first. This decision is already an environmental act, and it is therefore not surprising that people with a high environmental attitude tend to visit such events (Curtis, 2009; Nurmis, 2016). Also, paintings presented inside a museum are by nature less disruptive, as compared to, for instance, a painting somewhere outside (e.g., in the form of street art) or an installation or performance that can potentially physically block a path or take up a lot of space. Even though a painting can trigger an emotional reaction and a thought process that can deeply move someone, its physical representation in itself is less disruptive, at least as long as it stays in its framed context on a wall inside, hence, it does not impose in the same way.

Three outstanding artists and their works that are both high on the political engagement dimension and the level of disruptiveness are Joseph Beuys' "7000 Oaks," Olafur Eliasson's "Ice Watch," and Michael Pinsky's "Pollution Pods." The following sections present these artists and their work, followed by a review of research from art perception and environmental psychology and the effect of emotions on engagement.

3.2 Joseph Beuys: 7000 Oaks

One of the most famous installations by the influential German artist Joseph Beuys (1921–1986) is "7000 Oaks." For Documenta 7, an art fair in Germany (in 1982), Beuys prepared 7000 oak settlings together with the same number of basalt stones to be planted and placed all over the city of Kassel (Körner & Bellin-Harder, 2009). First, he placed the settlings and stones in an hourglass shape in front of the Documenta building. The hourglass shape was a metaphor – chosen to represent time and the progress of the installation. Then, the citizens of Kassel were invited to contribute to the "afforestation" of their city, by taking a tree from the hourglass and planting it somewhere in Kassel, placing the basalt stone just besides the tree. Beuys goal was to change the face of the city that had been dominated by automobile infrastructure and concrete. Also, he wanted to create awareness and guide a social transformation, all along with a piece of art that would last around 300 years (the normal life span of an oak tree; Körner & Bellin-Harder, 2009).

For Beuys, social transformation and art went hand in hand. He was convinced that everything, even "fat" or "speech," was a material that could be used creatively and turned into art, as long as it conveyed meaning (Beuys, 2007, p. 56). Beuys developed a theoretical framework for his creative practice and therein described the term of "Social Sculpture." For him, a "Social Sculpture" was the final or never-ending sculpture that leads to a transformed society. The transformation should be achieved through a democratic and participatory process (Adams, 1992) and by including citizens into the realization of his installations, as in "7000 Oaks."

"7000 Oaks" can be described as a disruptive piece of art for several reasons. First, because it actually changed the cityscape of Kassel; second, because it interrupted people's day-to-day practices and asked them to get involved to make the installation happen; and, third, because Beuys' intention was to create social, political, and practical disruption through this piece. Therefore, he chose the city of Kassel instead of a location outside of the city for his "afforestation." The idea to plant trees inside the city was not well received by the city council and the fact that technical infrastructure immediately got in the way underlined that (Körner & Bellin-Harder, 2009). Questions, such as "How do you plant a tree in a densely paved city?" or "What is more important to us, streets on which we can drive our cars, or fresh air from the trees?" were instantly prompted through the planting of the trees, displaying the social criticism of the art and sparking political discussions (Körner & Bellin-Harder, 2009). In fact, the process of planting the trees and bringing the artwork to a finish was so difficult and disruptive that it took 5 years to plant all the trees.[1] To facilitate the process, two of Beuys' assistants had to help throughout the years with the approval process and other problems. Moreover, "7000 Oaks" was an expensive endeavor, since apart from providing trees and the basalt and making the changes in the infrastructure, the trees had to be maintained after being planted. To raise money, Beuys did all kinds of things, including printing and selling flyers, as well as making an advertisement for a Japanese whiskey,[2] only under the condition that the TV commercial would read that all of the money was going to his project "7000 Oaks."

However, Beuys considered the debate and the challenges as part of process of the "Social Sculpture." The impact the installation supports that it was worth the effort: Körner and Bellin-Harder (2009) claim that "7000 Oaks" consequently affected the way trees are being handled and maintained in German urban landscape architecture. The installation was furthermore described as the symbol for the urban greening movement of the 1970s and 1980s (Grande, 2009) and to have made ecological aspects in landscape architecture mainstream (Barton, 2020). On top of that, the installation has become an important part of the cultural heritage and

[1] https://www.tate.org.uk/art/artworks/beuys-7000-oak-trees-ar00745

[2] https://www.youtube.com/watch?v=ARS3TO9r_z4

infrastructure of Kassel and has been awarded the status of a monument (Körner & Bellin-Harder, 2009).

3.3 Olafur Eliasson: Ice Watch

More recently, since the mid-1990s,[3] the Icelandic-Danish artist Olafur Eliasson gained a lot of attention with his light installations and environmental installations. He is running a studio in Berlin, where his team is producing the metal fixtures and other parts for his eclectic and monumental installations. Eliasson is known for his belief and intention to contribute to positive societal and environmental changes through his art (Nurmis, 2016). To do so, he plays with the cultural meaning of climate and the bodily experience of aspects of climate and the environment. In an interview for his show "The Weather Project" at Tate modern in 2003, he outlined how much we humans are dependent on favorable weather conditions and how much effort we invest in controlling the climate in our build environment, be it through heating or air-conditioning units.[4] However, outside of our buildings, we are totally exposed to the whims of the weather. Through his work, he invites us to reflect on this human conundrum. Offering to people a space for such reflections opens the possibility up for them to connect with the topic on their own terms. Recent studies have shown that such personal reflections are important to create engagement (Klöckner & Sommer, 2021; Sommer et al., 2019; Sommer, 2020; Sommer & Klöckner, 2021).

One of Eliasson's works, specifically related to climate change, is the installation "Ice Watch," that is part of his bigger work on melting ice and glaciers called "Your waste of time."[5] The installation "Ice Watch" was shown for the first time during the UN climate change negotiations, COP20 in Copenhagen, second time during COP21 in Paris, and third time in 2018 in front of Tate Modern in London. All three times, several pieces of ice from the Nuup Kangerlua Fjord in Greenland were transported to the locations in the cities, in order to let citizens experience the melting of the Antarctic ice caps. By doing so, Eliasson highlights the transience of this 2500-year-old ice that is melting at a rapid pace with current levels of climate change. Eliasson wants the spectators to expand their frame of reference, to bodily experience the chill of the ice and its crackling while it is melting, and to sense the time we are wasting.

"Ice Watch" qualifies as disruptive and engaging art because the ice blocks can be touched, smelled, listened to, and even crawled upon, giving the visitors an all-round bodily experience that leaves them with awe. The icebergs are made from compressed snow and contain bubbles of air that is hundreds of years old,

---

[3] https://olafureliasson.net/biography

[4] https://www.tate.org.uk/context-comment/articles/the-weather-project

[5] https://olafureliasson.net/

containing much less CO_2 and pollutants than air now. Minik Rosing, the scientist on the project states that the melted ice is so pure that it could even be drunk. While Eliasson adds that the possibility to touch and feel the ice makes "the knowledge in our head" a "knowledge of our hands,"[6] referring to the emotional connection "Ice Watch" is creating. Adding to the statement of impact the installation is creating are reviews from many art magazines and newspapers.[7,8,9] The reviews characterize the experience as "physical and immersive," "immediately striking," and as turning statistics on how fast the arctic ice shield is melting into a personal experience.[10] The Standard even declares the art "a spectacular call for action,"[11] supporting Eliasson's and Rosing's hope to have inspired visitors and decision-makers to take steps to reduce greenhouse gas emissions.

3.4 Michael Pinsky: Pollution Pods

Michael Pinsky is a UK-based artist, who focuses in his work on environmental issues and societal phenomena. Oftentimes, he challenges conventions, such as the British habit to eat food from chain restaurants such as Pret-a-manger ("Pret-a-manger"[12]) or commuting practices ("In Transit"[13]). To do so, he innovatively combines the roles of artist, activist, urban planner, and citizen. Pinsky is also inspired by the international locations in which his work will be exhibited, where he uses the social, physical, and political environment as resources.

Pinsky was the commissioned artist within the Climart project, collaborating with environmental psychologists and natural scientists to create a piece of art that engages the spectators with the topic of climate change. He shaped "Pollution Pods," taking the findings from early studies within Climart that found that an emotional and cognitive activation is key for individual's engagement and the discussion within the team into consideration (Roosen et al., 2018). "Pollution Pods" consists of "[F]ive interconnected geodesic domes that contain carefully mixed recipes emulating the relative presence of ozone, particulate matter, nitrogen dioxide, sulphur dioxide and carbon monoxide which pollute London, New Delhi, San Paolo and

---

[6] https://www.theguardian.com/artanddesign/2018/dec/11/icebergs-ahead-olafur-eliasson-brings-the-frozen-fjord-to-britain-ice-watch-london-climate-change

[7] https://publicdelivery.org/olafur-eliasson-ice-watch/

[8] https://www.artlyst.com/news/olafur-eliasson-ice-watch-installation-highlights-climate-change-in-london/

[9] https://thevcs.org/index.php/gardened-god/melting-our-eyes

[10] https://www.riseart.com/guide/2280/art-exhibitions-olafur-eliasson-s-ice-watch-at-tate-modern-rise-art-review

[11] https://www.standard.co.uk/culture/olafur-eliasson-and-minik-rosings-ice-watch-review-big-freeze-comes-early-with-an-urgent-warning-of-climate-crisis-a4015216.html

[12] http://www.michaelpinsky.com/portfolio/3854/

[13] http://www.michaelpinsky.com/portfolio/in-transit/

Beijing. Starting from a coastal location in Norway, the visitor passes through increasingly polluted cells, from dry and cold locations to hot and humid."[14] By walking through the "Pollution Pods," the visitor gets a bodily experience of what it means to live in environments with different levels of pollution from all around the world. It also highlights the connection between the capitalist consumerism in the Global North and the impact this has for millions of people mostly in the Global South.[15]

Given that the artwork was created with the clear intention to engage its audience, the political thought behind it is evident. Moreover, the smells, temperature differences, and varying levels of humidity within the domes provide an immersive experience for the visitors that disrupts their everyday routine. For this reason, **Pollution Pods** came in 71st place in the online art magazine "Artnet" list of the *100 Works of Art That Defined the Decade,*[16] and BBC Culture named the "Pollution Pods" "as one of the best ways to change the world."[17] Finally, the size of the installation and the fact that it was mostly exhibited outside and therefor easily accessible to the public adds to its disruptive nature.

3.5 Psychological Reactions to Disruptive Environmental Art

Theory from art perception and environmental psychology tells us that two categories of reactions to environmental art are especially important, when it comes to creating engagement through art:

1. Emotions, which fulfill a "relevance function," basically stressing why the experienced art is relevant for the person right now (Pfister & Böhm, 2008; Pelowski et al., 2016).
2. Thought processes around the meaning of the artwork, called "meaning-making" or "cognitive mastery" (Pelowski et al., 2016).

Theory suggests that both the emotional and cognitive reactions to the art are highly subjective processes (Pelowski et al., 2016). This is the case since they are not solely based on the objective representation of the art but blend it with personal experiences and memory. The subjectivity makes it on the one hand difficult for researchers to identify common mechanism that applies to all humans. On the other hand, the freedom of interpretation and the fact that art invites the individual to reflect and discover the meaning of the art for themselves are powerful. This personal space of reflection is what differentiates art from other environmental

---

[14] http://www.michaelpinsky.com/portfolio/pollution-pods-2/

[15] https://www.climart.info/pollutionpods

[16] https://news.artnet.com/opinion/100-works-that-defined-the-decade-part-2-1738773

[17] https://www.bbc.com/culture/article/20200422-climate-change-art-the-best-ways-to-change-the-world

information and awareness campaigns (Sommer, 2020). Anyone interested in the cognitions identified to be important in environmental art perception process is referred to publications of the Climart project (Sommer & Klöckner, 2021; Sommer et al., 2019; Sommer, 2020; Klöckner & Sommer, 2021). The focus of the rest of the chapter will be on emotions as they are particularly relevant with respect to disruptive communication.

3.6 Emotional Reactions to Environmental Art

Why are emotional reactions important when it comes to creating engagement with environmental art? One way of thinking is that the artwork is a medium with which artists transmit their emotions and thoughts. Consequently, if the viewer of the art is able to pick up on these emotions and discover the meaning the artist intended to show, the art is appraised as good and meaningful (Pelowski et al., 2020).

More generally, an artwork that moves us makes us feel something. This is also true for controversial or disruptive art. However, oftentimes in research on art perception, affective responses to art are only measured in a dichotomous way as liking vs. disliking the art. Silvia (2005, 2009, 2013) broke with this tradition, because he felt that a lot more could be learned about emotional reactions to art if modern theories from emotion psychology were applied in art perception. He described emotions as sophisticated, affective responses to art and applied the appraisal theory of emotions, to investigate the effect of art on several emotions including rejection (Silvia, 2005, 2009; Silvia & Brown, 2007; Cooper & Silvia, 2009).

The appraisal theory is useful especially in the context of environmental art because it is based on people's *evaluation* of a certain event, the experience and how this relates to a person's values, interests, and goals and not on the happening *itself*. It also means that the art can inspire someone to act, even if the artist had no environmental intentions when creating the piece. This makes art in itself is valuable, even if it is not expected to address a societal dilemma. Another upside is that through the connection to values, it easily relates to theory from environmental psychology that tells us that personal values are an influential factor for motivating environmental behavior (Stern et al., 1999; De Groot & Steg, 2007, 2008).

3.7 Hostile Emotions

In a study from 2007, Silvia and Brown investigated the specific effect of controversial art on negative emotions, such as anger and disgust. Implicitly, through appraisal theory, the study also looked at values participants were holding. Silvia and Brown (2007) found that anger was triggered by a controversial piece of art if the viewer perceived it to be incongruent with one's values and as intentionally offensive. If the art was perceived to be incongruent with one's values and not intentionally offensive but unpleasant, disgust was the ensuing emotion. In a follow-up study, anger and disgust predicted people's repressive attitudes against an image and whether they declined or accepted a free postcard of an offensive artwork (Cooper & Silvia, 2009). Unfortunately, the study does not address what happens if the art is perceived congruent with one's values but at the same time as intentionally offensive/disruptive and how this emotion would be associated with pro-environmental engagement. It can be assumed that a provocative artwork which initially disrupts your way of being (and potentially irritates) can become a powerful driver, once the meaning-making cognitions have established the congruence of values.

Connecting Silvia and colleagues finding on "hostile emotions" with findings from Climart, we find that the emotions "anger, sadness, and helplessness" together were the strongest emotional factor that increased visitor's intention to act on climate change and air pollution after the visit of the "Pollution Pods" (Sommer et al., 2019). The second strongest emotions were "a sense of awe, inspiration, and surprise," followed by "guilt and shame." Only one emotional factor negatively influenced changes in intentions: the factor "happiness and pride." According to the findings on anger mentioned above, one would expect that the "Pollution Pods" triggered anger by making people think of their values and feel that someone is intentionally not acting according with those values. For example, participants might have felt angry about people's lack of action to reduce air pollution. In that aspect, also the emotions "sadness" and "helplessness" seem to fit in. However, our initial assumption was that "sadness" and "helplessness" would inhibit pro-environmental intentions and behavior, as these are emotions which do not prompt action, but the results from our analyses suggest that this was not the case here.

3.8 Knowledge Emotions

The second type of emotions sparked by the "Pollution Pods" – "a sense of awe, inspiration, and surprise" – can be categorized together with the "knowledge emotions" described by Silvia (2009). The classical knowledge emotions "surprise," "confusion," and "interest" are related to the motivation to learn and to search for solutions. With respect to disruptive communication through art, this type of emotions appears to be crucial as it stimulates to engage more with the artwork and its meaning. Therefore, it can be concluded that effective disruptive artworks manage to create both "anger" and "awe," fuelling engagement through this mix of emotions. Motivation to learn about and find solutions to environmental problems is important when it comes to engaging people for a green cause.

Silvia (2009) describes "surprise" to be the initially strongest of the three emotions, because it interrupts people in their current action and orients them toward a possibly important new happening. Its main appraisal is that there is something unexpected and new. In the cluster analysis study on the ArtCOP21 data from Climart (Sommer & Klöckner, 2021), we found that participants perceive art that was rated highly "surprising" also to be "unusual" and that it "made them stop" in their tracks. This is in line with elements of the definition of disruptive communication given in Chap. 1. Surprising art also was reported to be challenging social norms and to be confrontational, reflecting other characteristics of disruptive communication. Moreover, it made participants aware of the impact of their actions and highlighted the relevance of climate change on their daily life. The "awareness of impact" and "relevance for daily life" were also found to be associated with climate change policy support (Sommer & Klöckner, 2021) and increased intentions to act on climate change and air pollution (Sommer et al., 2019). These associations between emotions and cognitive processes are part of the psychological mechanism with which disruptive environmental art can lead to engagement in individuals (Sommer, 2020).

Silvia (2009) describes the emotion "surprise" to be followed by either "interest" or "confusion." Interest is an emotion defined by an appraisal of high novelty and complexity, combined with high coping potential (Silvia, 2009). Confusion results from the appraisal of the event as new and complex but also as difficult to deal with (Fayn et al., 2019). This difference can be very relevant in the follow-up of a disruptive communication event. Making sure the initial emotional reaction is translated to interest rather than confusion seems to be of essence.

3.9 Self-conscious Emotions

Having covered "hostile emotions" and "knowledge emotions," another important group of emotions in the context of art is "self-conscious emotions" (Silvia, 2009). "Self-conscious emotions" are emotions like "shame," "pride," "regret," "embarrassment," and "guilt." These emotions are unusually complex, because a living being needs to have a sense of self and be able to reflect on it to feel these emotions. Therefore, many animals do not show these emotions and humans develop them relatively late in their childhood (Tangney, 2012). Silvia (2009) suggests that from an appraisal point of view, self-conscious emotions are triggered when consistencies or inconsistencies with cultural standards occur. Besides, "shame," "guilt," "embarrassment," or "pride" can be experienced as collective self-conscious emotions, meaning that we can feel them for people that we perceive to be similar to us – our in-group.

Looking into the field of environmental psychology and research on self-conscious emotions, we find that "guilt" is the one most widely researched in the context of environmental action. "Guilt" has been found to be associated with pro-environmental behavior (Mallett, 2012; Rees et al., 2015), also in the case of collective guilt (Ferguson & Branscombe, 2010). However, a study by Schneider et al. (2017) found an indication for a stronger effect of the positive valanced emotion "pride," in comparison to the negative valanced "guilt." Results show that anticipating "pride" from acting green as compared to anticipating feeling "guilty" from not acting increases pro-environmental behavioral intentions more. Contrary to these findings, in Climart we found that "guilt and shame" together were associated with an *increase* in pro-environmental intentions, while "happiness and pride" were associated to a *decrease*. This different effect can most likely be explained by the fact that data collection around the "Pollution Pods" took place in Trondheim and London and both locations were represented in the first two domes. These two locations exhibit significantly lower levels of air pollution than the following domes, which could have led visitors to feel "happy" not to live in the other places and "proud" that their cities/countries were environmentally speaking doing better than the others. This finding also highlights the importance of seeing emotions in the context they are being perceived.

As mentioned above, self-conscious emotions are mostly relevant in social settings. Therefore, including framing messages referring to the in-group as opposed to an out-group was found to influence the emotional reactions to the environmental message (Harth et al., 2013). If the in-group was framed as responsible for environmental damages, participants felt "guilty" and "angry." The intentions that followed the emotions, though, were different for the two emotions: feeling "guilty" led participants to report intentions to repair the environmental damages, while "anger" made them intent to punish the environmental sinners. This is an important distinction also relevant for disruptive artworks. If the art is able to induce guilt, more individual behavior change of the recipient can be expected than if it induces anger (where the action would be directed at others, e.g., politicians, the industry, or irresponsible people not acting). Lastly, if the in-group was framed as responsible for environmental protection, participants were feeling "proud" to be able to act and reported a higher intention to engage in environmental behaviors that favor the in-group.

3.10 Worry and Fear

Emotions not investigated so far in art perception but in environmental psychology as facilitators or barriers of environmental engagement are "worry" (Smith & Leiserowitz, 2014; Bouman et al., 2020) and "fear" (O'Neill & Nicholson-Cole, 2009; Skurka et al., 2018) In the following, they are discussed in relation to their potential role in the effects of disruptive artwork.

"Worry" is an emotion that is characterized as less of a calculated assessment than "perceived risk" and not as overwhelming as "fear" (Bouman et al., 2020). This makes "worry" according to Bouman and colleagues a relatively personal and dynamic emotion. It was also found by Smith and Leiserowitz (2014) to be the strongest predictor of climate change policy support, in comparison to other emotions, such as "fear," "interest," and "anger," as well as values and sociodemographics. Building on these and other findings, Bouman et al. (2020) assumed that "worry" would be particularly likely to influence pro-environmental behavior. Moreover, they expected that people with high biospheric values (which means that they value nature for itself, not for serving humans) should worry more about climate change. This "worry" would directly and indirectly impact climate policy support and personal climate mitigation actions via people's perceived "responsibility to act." The results of the empirical study that included data from 23 different European countries (Bouman et al., 2020) partly supported their hypotheses: people with stronger biospheric values indeed worried more. "Worry" was also associated with climate policy support and personal mitigation action, but the effect was mediated via personal responsibility and not directly. For disruptive art this could mean that creating worry instead of fear could be a promising avenue, meaning that the art should make people aware of the impact climate change will have on them personally, their families, or other beings they value highly, but not to a degree where it frightens them so much that they shut down emotionally. This should be rather difficult to achieve, given that people have very different levels of what frightens them.

More research pointing toward to the importance of differentiating between worry and fear is coming from health psychology, where scare tactics have been used to make people stop unhealthy behavior (smoking, unhealthy eating) or engage in healthy behavior (sports, safe sex practices). Results from two meta-analyses show that fear can be overwhelming and hinder wished-for behavior more than it facilitates (Peters et al., 2013; Kok et al., 2018). This is particularly true when combined with a message highlighting the personal losses connected to climate change (O'Neill & Nicholson-Cole, 2009). Often, when fear is induced, people have very low self-efficacy, meaning that they do not know what to effectively do to reduce the health risk (Peters et al., 2013) or environmental problem, so that they tend to avoid the topic instead of engaging with it (Kok et al., 2018). Artworks should, thus, restrain from inducing fear only.

However, one potential path of successfully using fear to motivate people to act on environmental topics was suggested by Nabi et al. (2018). Nabi and colleagues tested a theory of "emotional flow," with the hypothesis that the succession of "fear" by "hope" would make the feeling of "hope" stronger, mostly because "fear" physically arouses the body and therefore makes the recipient of the message more susceptible to the following emotions. To test this hypothesis, "fear" was induced through a newspaper article, incorporating a threatening climate change story. Then either "fear" or "hope" was triggered, using a loss-frame ("Not Stopping Climate Change Threatens US Health") and a gain-frame ("Stopping Climate Change Will Benefit US Health") respectively. Results show that the emotion "hope" is a key mediator for climate change attitudes and support. That means that if fear is followed by hope, it positively affects attitudes and supports for climate change policies. This paves the way for assuming that disruptive art that first triggers fear but then gives spectators hope might be of particular power. Speaking of power and potential impact of disruptive environmental art, emotional reactions to disruptive art are not just relevant for motivating citizens to take action in their daily lives but can be especially effectful when they are experienced by policy- and other decision-makers. The "Pollution Pods" have toured the world and were among other locations also exhibited at the UN Youth Climate Summit in New York (2019) and COP25 in Madrid. At both exhibitions, the installation was visited by leaders working for the UN, WHO, and Heads of Nations. If these politicians felt emotionally moved through the "Pollution Pods," this emotion might have turned into fuel to push an environmental political agenda further and potentially translate into the fundamental change we need. This is true for any piece of disruptive art that interrupts the daily routine of the general population but also politicians and other stakeholders.

3.11 Summary

To summarize, disruptive environmental art such as "7000 Oaks" has proven to be able to change the cityscape of Kassel but also how ecology and landscape architecture are understood and applied in city-planning. "Ice Watch" and "Pollution Pods" have according to the media reports and empirical studies proven to affect people on an emotional level and to make climate change personally relevant. Theories and empirical findings from art perception and environmental psychology indicate why this is the case and how disruptive art can engage its audience speaking to emotions. The potential for disruptive environmental art is increased if the audience includes policymakers that can directly affect climate change mitigation and adaptation plans. Finally, emotions play a key role in motivating people to act. However, the herein presented studies show that emotions are not just levers that can or should be pulled to result in a specific intention or a certain wished-for behavior. Emotions have to be seen and understood in the social and cultural context the recipient is situated in and in interaction with other emotions.

References

Adams, D. (1992). Joseph Beuys: Pioneer of a radical ecology. *Art Journal, 51*(2), 26–34.

Barton, P. (2020). *From Land Art to Eco Art: An exploration of Joseph Beuys' pivotal social sculpture, 7000 Oaks – City Forestation instead of City Administration as a bridge between the two to inform the author's developing practice*. Leonardo.

Belfiore, E., & Bennett, O. (2007). Rethinking the social impacts of the arts. *International Journal of Cultural Policy, 13*(2), 135–151.

Beuys, J. (2007). *What is art?: Conversation with Joseph Beuys*. Clairview Books.

Boettger, S. (2002). *Earthworks: Art and the landscape of the sixties*. University of California Press.

Bouman, T., Verschoor, M., Albers, C. J., Böhm, G., Fisher, S. D., Poortinga, W., et al. (2020). When worry about climate change leads to climate action: How values, worry and personal responsibility relate to various climate actions. *Global Environmental Change, 62*, 102061.

Cooper, J. M., & Silvia, P. J. (2009). Opposing art: Rejection as an action tendency of hostile aesthetic emotions. *Empirical Studies of the Arts, 27*(1), 109–126.

Curtis, D. J. (2009). Creating inspiration: The role of the arts in creating empathy for ecological restoration. *Ecological Management & Restoration, 10*(3), 174–184.

De Groot, J. I., & Steg, L. (2007). Value orientations and environmental beliefs in five countries: Validity of an instrument to measure egoistic, altruistic and biospheric value orientations. *Journal of Cross-Cultural Psychology, 38*(3), 318–332.

De Groot, J. I., & Steg, L. (2008). Value orientations to explain beliefs related to environmental significant behaviour: How to measure egoistic, altruistic, and biospheric value orientations. *Environment and Behaviour, 40*(3), 330–354.

Demos, T. J. (2016). Between rebel creativity and reification: For and against visual activism. *Journal of Visual Culture, 15*(1), 85–102.

Fayn, K., Silvia, P. J., Dejonckheere, E., Verdonck, S., & Kuppens, P. (2019). Confused or curious? Openness/intellect predicts more positive interest-confusion relations. *Journal of Personality and Social Psychology, 117*(5), 1016–1033.

Ferguson, M. A., & Branscombe, N. R. (2010). Collective guilt mediates the effect of beliefs about global warming on willingness to engage in mitigation behaviour. *Journal of Environmental Psychology, 30*(2), 135–142.

Grande, J. (2009). The lay of land art: Out of museum+ onto the earth. *Public Art Review, 20*(2), 42–49.

Harth, N. S., Leach, C. W., & Kessler, T. (2013). Guilt, anger, and pride about in-group environmental behaviour: Different emotions predict distinct intentions. *Journal of Environmental Psychology, 34*, 18–26.

Klöckner, C. A., & Sommer, L. K. (2021). Visual art inspired by climate change—An analysis of audience reactions to 37 artworks presented during 21st UN climate summit in Paris. *PLoS One, 16*(2), e0247331.

Kok, G., Peters, G. J. Y., Kessels, L. T., Ten Hoor, G. A., & Ruiter, R. A. (2018). Ignoring theory and misinterpreting evidence: The false belief in fear appeals. *Health Psychology Review, 12*(2), 111–125.

Körner, S., & Bellin-Harder, F. (2009). The 7000 Eichen of Joseph Beuys–experiences after twenty-five years. *Journal of Landscape Architecture, 4*(2), 6–19.

Mallett, R. K. (2012). Eco-guilt motivates eco-friendly behaviour. *Ecopsychology, 4*(3), 223–231.

Marks, M., Chandler, L., & Baldwin, C. (2017). Environmental art as an innovative medium for environmental education in biosphere reserves. *Environmental Education Research, 23*(9), 1307–1321.

Martin, J. (2015). *Ecocritical art in times of climate change: Tracing ecological relationships between humans and nonhumans through the hyperextension of objects* (Doctoral dissertation). Goldsmiths College, University of London.

Nabi, R. L., Gustafson, A., & Jensen, R. (2018). Framing climate change: Exploring the role of emotion in generating advocacy behaviour. *Science Communication, 40*(4), 442–468.

Nurmis, J. (2016). Visual climate change art 2005–2015: Discourse and practice. *Wiley Interdisciplinary Reviews: Climate Change, 7*(4), 501–516.

O'Neill, S., & Nicholson-Cole, S. (2009). "Fear won't do it" promoting positive engagement with climate change through visual and iconic representations. *Science Communication, 30*(3), 355–379.

Pelowski, M., Markey, P. S., Lauring, J. O., & Leder, H. (2016). Visualizing the impact of art: An update and comparison of current psychological models of art experience. *Frontiers in Human Neuroscience, 10*, 160.

Pelowski, M., Specker, E., Gerger, G., Leder, H., & Weingarden, L. S. (2020). Do you feel like I do? A study of spontaneous and deliberate emotion sharing and understanding between artists and perceivers of installation art. *Psychology of Aesthetics, Creativity, and the Arts, 14*(3), 276–293.

Peters, G. J. Y., Ruiter, R. A., & Kok, G. (2013). Threatening communication: A critical reanalysis and a revised meta-analytic test of fear appeal theory. *Health Psychology Review, 7*(sup1), S8–S31.

Pfister, H. R., & Böhm, G. (2008). The multiplicity of emotions: A framework of emotional functions in decision making. *Judgment and Decision making, 3*(1), 5–17.

Rees, J. H., Klug, S., & Bamberg, S. (2015). Guilty conscience: Motivating pro-environmental behaviour by inducing negative moral emotions. *Climatic Change, 130*(3), 439–452.

Roosen, L. J., Klöckner, C. A., & Swim, J. K. (2018). Visual art as a way to communicate climate change: A psychological perspective on climate change–related art. *World Art, 8*(1), 85–110.

Schneider, C. R., Zaval, L., Weber, E. U., & Markowitz, E. M. (2017). The influence of anticipated pride and guilt on pro-environmental decision making. *PLoS One, 12*(11), e0188781.

Silvia, P. J. (2005). Emotional responses to art: From collation and arousal to cognition and emotion. *Review of General Psychology, 9*(4), 342–357.

Silvia, P. J. (2009). Looking past pleasure: Anger, confusion, disgust, pride, surprise, and other unusual aesthetic emotions. *Psychology of Aesthetics, Creativity, and the Arts, 3*(1), 48–51.

Silvia, P. J. (2013). Interested experts, confused novices: Art expertise and the knowledge emotions. *Empirical Studies of the Arts, 31*(1), 107–115.

Silvia, P. J., & Brown, E. M. (2007). Anger, disgust, and the negative aesthetic emotions: Expanding an appraisal model of aesthetic experience. *Psychology of Aesthetics, Creativity, and the Arts, 1*(2), 100–106.

Skurka, C., Niederdeppe, J., Romero-Canyas, R., & Acup, D. (2018). Pathways of influence in emotional appeals: Benefits and tradeoffs of using fear or humor to promote climate change-related intentions and risk perceptions. *Journal of Communication, 68*(1), 169–193.

Smith, N., & Leiserowitz, A. (2014). The role of emotion in global warming policy support and opposition. *Risk Analysis, 34*(5), 937–948.

Sommer, L. K. (2020). *"EARTH" without "ART" is just "EH" An environmental psychological approach on climate change communication through art* (Doctoral Thesis). Norwegian University of Science and Technology.

Sommer, L. K., & Klöckner, C. A. (2021). Does activist art have the capacity to raise awareness in audiences?—A study on climate change art at the ArtCOP21 event in Paris. *Psychology of Aesthetics, Creativity, and the Arts, 15*(1), 60.

Sommer, L. K., Swim, J. K., Keller, A., & Klöckner, C. A. (2019). "Pollution Pods": The merging of art and psychology to engage the public in climate change. *Global Environmental Change, 59*, 101992.

Stern, P. C., Dietz, T., Abel, T., Guagnano, G. A., & Kalof, L. (1999). A value-belief-norm theory of support for social movements: The case of environmentalism. *Human Ecology Review, 6*, 81–97.

Tangney, J. P. (2012). Self-conscious emotions. In M. R. Leary & J. P. Tangney (Eds.), *Handbook of self and identity* (pp. 446–478). The Guilford Press.

Wallen, R. (2012). Ecological art: A call for visionary intervention in a time of crisis. *Leonardo, 45*(3), 234–242.

Wolfe, D., & Russell, K. (2010). Garden/art: The nature-based sculpture program of the South Carolina Botanical Garden. *Environmental Communication, 4*(2), 237–249.

Chapter 4
A License to Disrupt? Artistic Activism in Environmental Public Dissent and Protest

Berend Barkela, Teresa Gil López, and Christian A. Klöckner

4.1 Introduction

One of the most visible representations of public engagement with global environmental challenges is the climate movements that have recently emerged around the world calling for more ambitious political responses to climate change (Caniglia et al., 2015). The People's Climate Movement or Fridays for future (inspired by Greta Thunberg's School Strike) are major social movements that have realized large-scale demonstrations (Alkaher, 2020), though put slightly on hold by the COVID-19 pandemic and measures to contain it. These movements have made considerable achievements, especially in getting substantial media attention to introduce the idea of climate mitigation into the public agenda (von Zabern & Tulloch, 2021). Media attention is crucial to the success of such social movements. Greater attention on a discursive level, which can be triggered through protest, is helpful in advancing legislation to mitigate climate change (Nash & Steurer, 2021; Otto et al., 2020). However, such attention is not to be taken for granted. Many social movements are not represented in the media at all, and often, social movement protest is presented rather critically in mainstream media (McLeod, 2007). These critical portrayals include aspects as varied as arrest counts, justifications of police action, descriptions of violent actions from drum-playing, bandana-wearing protesters, or witness accounts used to make generalizations about public opinion (e.g., Gil-Lopez, 2021; Dardis, 2006; Lee, 2014).

Supporters and organizers of environmental movements often express disappointment that despite high attention, their efficacy is limited in terms of policy impact (de Moor, 2017). In such situations, when social movements perceive little efficacy in achieving their goals, they tend to move away from moderate actions and engage in more disruptive and even violent protest tactics (Tausch et al., 2011). So far, environmental movements have remained largely peaceful – although there are examples of extreme forms of civil disobedience (e.g., parliament occupation,

C. A. Klöckner, E. Löfström, *Disruptive Environmental Communication*, Psychology and Our Planet, https://doi.org/10.1007/978-3-031-17165-9_4

clashes with police; Bugden, 2020). The degree of contention, disruptiveness, or transgression of the tactics used to protest is fundamental in defining the interactions between social movement actors, the media, and the public and therefore crucial in enabling the accomplishment of the movement's goals. The more disruptive, counter-normative, or harmful to others these tactics are, the more negatively they are portrayed in the media, which in turn leads to less effectiveness in terms of public support and appreciation of the movement's goals (Feinberg et al., 2020). At the same time, protest must be disruptive to some degree, as it is in the nature of social movements to publicly contest prevailing power structures (i.e., challenge status quo). In fact, protest, as a collective way of interfering with the ordinary course of life (see Sect. 1.7) to publicly communicate dissent, may be the distinctive feature separating social movements from the routine political actor (Taylor & Van Dyke, 2004).

In this relationship of necessary disruption, media coverage, and political as well as social impact, artistic features of protest can play a distinctive role. In addition to prescribing the degree of disruption, the tactical repertoire of movements also involves the strategic use of visual and performative elements of protest, for example, images, symbols, or choreography in public spaces (McGarry et al., 2020). The political voice of protestors is not only produced by vocalized claims or demands but also through protest aesthetics (Butler, 2015). Such artistic activism is here understood as an art form which takes place outside institutions like museums, has an emancipatory basis, and is aimed at solving social rather than artistic problems (Weibel, 2015). A variety of creative expressions is already present among environmental activists, ranging from creatively designed protest banners (Catanzaro & Collin, 2021) to performative actions. For example, the "red rebel brigade," a group of individuals in prominent red costumes who move following a specific choreography, has become a recurring element of environmental protests around the world (Lavender, 2019). Art and disruption are therefore not necessarily opposites. Artistic protest can appear rebellious, disobedient, or provocative but expressing "rebel creativity" (Demos, 2016). Such disruptive artistic activism as a protest tactic might enable special interpretations of disruption (see also Chap. 3, Activist Art). In Western democracies in particular, a guiding principle that has emerged from the Art of the Enlightenment is that the arts critically reflect society, question accepted beliefs, and engage with the political realm (Belfiore & Bennett, 2007; Bishop, 2006). In this regard, there is a social consensus that art is allowed to and should be disruptive; it is granted a broad understanding of freedom of expression within the context of artistic freedom (Alexander & Rueschemeyer, 2005). A unique impact of artistic interventions is found in their ability to generate and alter meanings in ways that disrupt habits and social norms and thus provoke reaction (Teune, 2005). Based on this notion, artistic forms of protest are expected to change the way disruption is perceived and interpreted by media and the public. Within the environmental movement, such an approach can be found, for example, in tactical considerations of activists involved in "Extinction Rebellion," who assume that performance art combined with civil disobedience is more effective than law-abiding marches in terms of media coverage and political discussion (Burgess & Read, 2020).

In this chapter, we will discuss the extent to which disruptive artistic action can be a reasonable protest tactic. Since we expect the media to play a crucial role in movement outcomes, we will relate such strategic considerations of social movements to journalistic and media routines and their impact on public opinion. In what follows, we will first address the question of how likely disruptive and artistic protest tactics are to be selected by media (public attention). Second, we will examine the ways in which these forms of protest can be expected to be represented and framed by media (public perception). Third, we question how these media portrayals can affect media recipients, that is, the public, policymakers, and the electorate (public opinion). Finally, we hypothesize an equilibrium within which disruptive artistic action is likely to be effective.

4.2 How Does Disruptive Artistic Activism Generate Public Attention?

On Monday, September 22, 2014, the New York Times – considered as one of the most influential newspapers worldwide – placed the People's Climate March on the front page (Foderaro, 2014). With four pictures, headline, and three columns of lead story, this event was not only selected for coverage but also prominently placed and given considerable space. Admittedly, a degree of media attention is more often an exception than a rule. As mass media have only limited space for news, protests compete with multiple other events. This creates a circumstance where many protests do not get into the news at all. When media coverage was checked against police registries of daily protest events, only about 10 percent of protest events, half at most, were seen reported on television on the same day or in newspapers the following day. Many of these are often not very prominently placed and rather short (Wouters, 2013).

However unfavorable this landscape may seem, movements continue to fight for media attention. It is primarily through the media that their messages can reach not only potential supporters but also the general public (Gamson, 2004). Without the public's attention, movements are powerless in their goal of influencing public opinion (Vliegenthart & Walgrave, 2012). Yet, because movements rarely possess the institutional power that other political actors have, their relevance for reporters is not automatically established: movements must generate newsworthiness (Vliegenthart & Walgrave, 2012). Generating newsworthiness often implies creating events with characteristics that are suitable and attractive for reporting in accordance with certain media logics. Events that conform to well-established news values such as conflict or surprise (Harcup & O'Neill, 2001) and that align with professional routines constraining the work of reporters (Shoemaker & Reese, 2014) have greater chance to get past the gatekeeper(s). Naturally, movements have devised protest tactics strategically aimed at attracting media attention (Taylor & Van Dyke, 2004). The most reliable and relevant characteristic is the size of a protest event: the larger the crowd a movement can motivate to participate, the more likely it is that the media will report on it at all and that such reports will be prominently placed and longer (Wouters, 2013).

The presence of both disruptive and artistic actions can also impact the amount of coverage, though perhaps to different degrees. First, disruption can range from civil disobedience to extreme violence, and its intensity is positively correlated to the amount of coverage (Myers & Caniglia, 2004). Aspects such as the presence of police, the number of arrests, and the amount of damage caused during protests are often given prominence not only because of their inherent conflict value but also because they are regarded as objective, quantifiable observations and are generally easy to obtain through official source channels (McLeod, 2007). Similarly, artistic protest actions possess characteristics that could contribute to increase media attention. This potential is especially attributed to their ability to create visual images (Doerr et al., 2013): visual content is not only expected to reach immediate public attention, but it also constitutes documentation through iconographic images, making the protest more durable in time (McGarry et al., 2020). The more creative a visual message, the more likely it is to be multiplied through social networks (Fichnová et al., 2019) and to achieve wider outreach and amplification (Boone et al., 2018). Other, more performative acts, such as theatrical enactments, are also able to raise awareness and increase media outreach (Sanz & Rodriguez-Labajos, 2021). Yet another characteristic is the participation of professional and recognized artists in social movements, which helps to create iconic symbols that achieve media attention (NicGhabhann, 2018). When quantifying these effects, the presence of symbolic actions (including colorful, vivid, or dramatic actions) in principle provides media attention, although it is less likely to achieve prominence or length in the way size and disruption do (Wouters, 2013).

Considering this and the fact that visual content is more likely to be selected by journalists if unexpected, surprising, or disruptive motives exist (Geise et al., 2021; Rössler et al., 2011), artistic actions may get more far-reaching attention when combined with disruption. This is suggested by some case studies of artistic protest events. One example is the "standing man" – an iconic performance of the Turkish artist Erdem Gündüz, who stood silently looking at a banner of Kemal Ataturk for 8 hours during the 2013 Gezi Park protests. This silent appeal for a secular Turkey attracted a lot of attention as well as imitators, while the parallel demonstrations were suppressed with brutal police violence. Such physical performance, interpreted as an act of defiant resistance (Ryan, 2019), illustrates how attention can be achieved through artistic disruption. A more extensive performance was investigated in a case study of environmental activists in Indonesia, who were protesting against the commercial development of a local city forest through theatrical enactment. Among several actions, activists staged a performance wherein mime dancers expressed their criticism and demands. This pantomime attracted the attention of journalists and bystanders, who recorded the action on their mobile devices. Not only was the event captured on private and media cameras, but reporters also interviewed activists after the performance and asked them to explain this unusual event; thus, their claims and demands were effectively transmitted (Alam & Nilan, 2018).

Another study investigated the disruptive performance of a Buenos Aires-based family, who staged their planned beach vacation inside a bank after finding out they no longer were able to access their savings due to a generalized government-led freeze on bank accounts during the economic crisis in 2001. After having informed local news stations, the performance was televised live, attracting more participants and media interest (Fuentes, 2012).

In these examples, activists achieved significant media attention for their issues through artistic actions while using moderate disruption – and having relatively few participants. The scenes illustrate how artistic strategies can make social movements more interesting to journalists. Although artistic protest actions alone could risk being placed less prominently, their combination with disruption – in principle more effective in attracting media attention – may even decrease the relative importance of a movement's size for getting into the news. Still, pictures of colorful performances and violent depictions can compete (Neumayer & Rossi, 2018), and images of violence seem more likely to meet the demand of news producers (Veneti, 2017).

4.3 How Is Disruptive Artistic Activism Covered by Mass Media?

While prominence and volume of media coverage bring social movements into the public agenda, the way in which content is presented further determines public perceptions about movements and their issues. As intermediaries between the movement and the public, the media contribute to define social movements, protests, and their underlying causes. Fundamental questions of definition and representation have traditionally been addressed by media and communication scholars through the theoretical prism of media framing (Entman, 1993). This approach considers not only the evaluation of news events, for example, in commentary or editorials but also the (un-)intended emphasis on some aspects and facets of an issue – in particular problem definitions, causal interpretations, evaluations, and remedies and responsibilities – over others (Entman, 1993).

Much like with media attention, the relationship between movements and media is seen as rather asymmetrical with regards to framing. On one hand, movements strive to receive coverage in line with their own frames. On the other, given the event orientation of news, mainstream media tend to focus their attention on single dramatic or spectacular protest events. As a result, they often fail to acknowledge the movement's claims and demands as well as to describe them as a political event (Gitlin, 1980; Smith et al., 2001). Furthermore, mainstream protest coverage has traditionally followed a stable pattern of representation which highlights the non-normative, violent, and deviant nature of protest groups. The "protest paradigm" (Chan & Lee, 1984; Gitlin, 1980) is assumed to reflect the media's role in supporting the status quo, which leads them to delegitimize social movements. The more movements challenge the status quo and the higher the extremity of their protest tactics (i.e., the extent to which they violate social norms), the more negatively the media will treat a movement (Shoemaker, 1984). Coverage will then focus on specific protest tactics and push actual claims and demands to the background (Boyle et al., 2012). Since the tactics that achieve high attention are the same ones that prompt more delegitimizing coverage, there is an "activist's dilemma" (Feinberg et al., 2020).

Finding the balance between the effectiveness of disruptive strategies and avoiding perceptions of radicality is therefore crucial. Artistic tactics are seen as an opportunity to avoid normative disapproval and therefore to receive a more positive evaluation (Branagan, 2003). The narrative possibilities of (utopian) artistic depictions offer constructive communication strategies and thus circumvent potential allegations that protest actions are a threat to society (Duncombe & Lambert, 2017). What is more, even in light of violent protest action, nonviolent protest groups can facilitate differentiation from radical elements by establishing visual and performative action (O'Neill, 2004). This demonstrates that artistic protest tactics do not necessarily have to be nondisruptive. On the contrary, the fundamental motivation of social movements to challenge the status quo shows certain parallels to art, mainly in the common ambition of promoting social legitimacy and acceptance for new ideas (Baumann, 2007). Given that the extent of negative coverage is related to the degree of disruption – ranging from disobedience to violence – artistic actions can have at least a moderating influence. When art and activism are seen as a socially engaged artistic practice of the radical avant-garde, they are effectively legitimized by the freedom of art (Grindon, 2011). Furthermore, in the tradition of political carnival, questioning existing power relations in a disrespectful and disruptive way, but with creative means, can offer a window of opportunity for progressive social change (Bruner, 2005).

However, creative and symbolic actions offer only a limited opportunity of escaping the protest paradigm, and artistic representations are not a guarantee of a positive media portrayal. A series of media frames have been defined in the literature which characterize negative coverage of artistic expressions in protests. McLeod and Hertog (1999) identified the *circus* frame among the five most common ways of representing radical protests in mainstream media coverage. This media frame generally highlights the spectacular, theatrical, or performative aspects of protests. It also emphasizes "atypical," symbolic behaviors of protesters, such as dancing and flag burning, in ways that ridicule them (Kilgo et al., 2018). Artistic expressions are thus described as amateurish or eccentric. The circus frame is also often referred to as carnival (Dardis, 2006), performance (McFarlane & Hay, 2003; Kleut & Milojevic, 2021), spectacle (Gitlin, 1980; Harlow & Johnson, 2011; Harlow et al., 2017), or show (Xu, 2013). Yet another similar frame is the freak frame, which is not unique to artistic performances but refers more generally to the odd or unconventional behaviors displayed by demonstrators – news stories tend to call attention to carnivalesque, dramaturgical performances but also to peculiar aspects of appearance such as piercings, dyed hair, or clothing (e.g., Ashley & Olson, 1998; Gitlin, 1980; McLeod & Hertog, 1999).

It is not always the case that associations of protests with metaphors of carnival or circus signify negative coverage. After all, these associations are not rare and relate to the essential use of art and symbolism in contemporary protest. As an illustration, findings are mixed with regard to whether show/carnival framings are invariably associated to the overall negative tone of coverage as much as violence and radicality. While Dardis (2006) did not find their presence to bear any relation to the overall tone of stories, Xu (2013) did. A plausible explanation for this could be that, unlike other predominant negative frames which are rooted in ideas of radicality or violence, what is unique about carnival and circus metaphors is that they tie performative and dramatic elements of protests to notions of eccentricity, madness, and ridicule (McFarlane & Hay, 2003). Accordingly, coverage may not necessarily be articulated in a negative tone nor convey it, but it can nevertheless be delegitimizing insofar as it portrays movements as trivial or futile (McLeod, 2007).

Hence, while artistic protest tactics may not ensure positive framings, they at least seem to ensure that movements are portrayed as less radical than when using other forms of disruption. Little research has looked at this interaction in detail, but some results indicate that artistic disruption is granted creative licenses which may foster a potential for disruption and transgression that other protest actions do not possess (NicGhabhann, 2018; O'Neill, 2004). For example, as a counterweight to the usually violent Black Bloc, a Pink Bloc has been established as a protest tactic, where protesters are dressed as fairies instead of wearing black hoods and throw teddy bears instead of rocks at the police. Such performances are a humorous and media-effective confrontation with the police that is difficult to criticize, while at the same time they counteract the tendency for reporters to portray a protest movement as violent (NicGhabhann, 2018; O'Neill, 2004). Similarly, during a climate protest 2019, organized by Extinction Rebellion activists, an inflatable octopus was paraded through London. There were several encounters with the police as activists attempted to control the octopus' direction. Instead of leading to violence, this resulted in rather absurd media images of policemen trying to keep a giant octopus at bay (Tunalı, 2020). When authorities are more hesitant to respond to protests, as is clear from the examples, creative images emerge which are more ambiguous than street battles. In staging scenes of historical revolutions during antigovernmental protests in Bulgaria in 2013, activists avoided governmental oppression while placing their message, visually at least, in national and international media (Trifonov, 2017). Staged and playful situations can thus "prank" media logics to transport a message, regardless of how media comment on the situation (Harold, 2004). Nevertheless, government agencies can also display authoritarian responses to artistic activism. Artistic provocations by groups like Voina and Pussy Riot have caused strong reactions by governments and were linked to the semantic field of terrorism in some strands of media coverage (Imposti, 2015). Several studies suggest that even the liberal-oppositional press in Russia, as well as the international media, failed to interpret the performance and highlighted the governments' response instead of the message intended by the artwork (Kananovich, 2016; Steinholt, 2013). However, the image's iconicity made them go around the world and brought solidarity, since the case was eventually also interpreted as an instance of art censorship (Bruce, 2015; Weij et al., 2015).

In sum, on the one hand, disruptive action, whether creative or not, can push movements' claims and demands into the background – protests become spectacular media events without political relevance (Robertson, 2006). On the other hand, it becomes clear that an aesthetic approach to disruption is able to irritate established structures while offering greater critical distance and more room for interpretation than violence against things or people (Stanisavljevic, 2016). Artistic strategies may not be absolute alternatives in the sense that they would generally be evaluated by the media as political and positive. Yet, differences in the framing of nonnormative disruption with or without artistic expression seem to differ in their consequences. While the former is likely to be interpreted as a serious societal threat, the latter is more often connected to ridicule, madness, or eccentricity. In that respect, artistic strategies account for less negative framing with regard to normative evaluations, while they still hold a clear potential to interrogate established norms and meanings. In addition, they provide room to create images in the media, which contain messages that can be interpreted not only as destructive.

4.4 How Does Disruptive Artistic Activism Influence Public Opinion?

Both the level of attention and the way social movement protest is portrayed through the media ultimately affect the impact movements can have in the public. Much of what the public will come to know about movements and their underlying causes, which institutions they target or the circumstances under which the protest event is carried out, will be largely shaped by the media discourse. Independent of the direct experience that each individual has with a particular group or cause, their understanding of the event will in many ways depend on what aspects news stories highlight or leave out. Given this, how the mainstream media chooses to represent dissent and what voices, views, and values they choose to echo are integral aspects of the politics surrounding social protest. In the mediatized relation between protest action and the public, it is in the strategic interest of social movements to win public support (by mobilizing supporters), reach out to political elites, and impact political opinion formation among citizens (Vliegenthart & Walgrave, 2012).

To date, the examination of the impacts of media portrayals on citizens' attitudes has mostly been carried out by scholarly works on psychological theories of social identity. This work shows that social movements are more effective in achieving their goals when they manage to generate a common identity within the movement and in the eye of the public (Simon & Klandermans, 2001; van Zomeren et al., 2008). The more individuals identify with a movement, the more willing they are to join protest events or related organizations (Stürmer & Simon, 2004). This mobilizing function of protest is important, since it fosters an increase in movement size, and this, in turn, heightens its relevance. While it is true that most people who perceive protest in public will not participate in the movement themselves, they may still sympathize with the movement and indirectly support the protestors or their stand on a particular issue (Feinberg et al., 2020). Besides, movements do not necessarily have to convince the (whole) public of their claims and demands. Especially in a democratic context, political change can be achieved not only when citizens endorse sustainable policies but also when they accept them (PytlikZillig et al., 2018). In addition, a legitimate public perception is a requirement for movements to reach out to policymakers, validating the issue not only in the public but also in the political agenda and legislative processes (Vliegenthart & Walgrave, 2012).

If disruption is present, protest may be interpreted as unworthy, which weakens public identification and support (Wouters, 2019). Research on the effects of mainstream protest coverage on public opinion has repeatedly shown that a focus on the disruptive aspects of protest in news stories activates stigmatizing cognitions and delegitimizes protesters in the public eye. As an illustration, the extent to which news stories portray protest actions as violent, criminal, or deviant has been shown to have a significant impact in perceptions of legitimacy of specific protest groups but also of social protest as a political tool in general (McLeod, 1995; McLeod & Detenber, 1999; Shoemaker, 1982). Furthermore, protest stories featuring contentious or violent actions can lead to reduced support and mobilization (Gutting, 2019), reduced identification with the group (Simpson et al., 2018), or lower perceptions of morality (Orazani & Leidner, 2019). Although these negative effects depend, in part, on whether observers endorse the cause (e.g., Hsiao & Radnitz, 2020), widespread public support is arguably more likely if protest is less disruptive. Violent action decreases perceptions of issue legitimacy and group efficacy even among sympathizers (Thomas & Louis, 2014). Even moderate degrees of disruption, such as civil disobedience, reduce support especially among independent bystanders (Bugden, 2020).

Results on the effects of media coverage on visual, creative, or artistic protest tactics are far less developed. With regard to visual representations, images of (peaceful) protest marches make people feel they can do something about climate change (Metag et al., 2016; O'Neill et al., 2013) and can evoke idealizations of democracy as protestors are seen exercising their right to express a point of view publicly (Krause & Bucy, 2018). However, compared to other visual contents, images of protest are often highly polarizing, considered a cliché in climate imagery, and not generally recommended in strategic climate communications (Wang et al., 2018). Considering artistic expression, various authors conclude that art in general has a key role in communicating sustainability by generating attention, evoking emotions, and in some cases even influencing environmentally friendly behavior (Blasch & Turner, 2016; Curtis et al., 2014; Sommer & Klöckner, 2021). With regard to its significance for social movements, arts have the potential to foster shared identities (Curtis, 2011; Kay, 2000), leading to the assumption that artistic actions could promote identification with social movements. Artistic protest action also contributes to overcoming demographic barriers and facilitates engagement (Sanz & Rodriguez-Labajos, 2021). Performance-based tactics provide symbols uniting diverse international movements, and, having a wide reach, they can further mobilize people as well as constituencies (O'Neill, 2004). Similarly, perception of artistic protest can foster a positive emotional connection to the issue and evoke a sense of change in the public (H. E. Ryan, 2015). Furthermore, emerging evidence from focus groups suggests that protest images using the carnival frame are assessed positively, create empathy for the protesters, and support for the cause (Kenix, 2008). Yet, other focus groups indicate that performative actions in an environmental context can be perceived as lunatic, hipsterism or "typical environmentalist" (Chapman et al., 2016). It has not yet been researched how these mixed evaluations arise. However, relevant factors could be that the studies' samples are not representative of the general population and that different environmental issues were investigated.

The extent to which media effects of disruption can be balanced when artistic action is present cannot yet be evaluated on the basis of specific scientific evidence. However, some expectations can be derived from what is known about art as it relates to human perception. In principle, art is processed by recipients as a unique artistic experience that differs from everyday experiences (Pelowski et al., 2017). An art context leads, for example, to a distanced mode of processing, including a more emotionally positive evaluation of the observed content (Cupchik, 1995). Even negatively valenced press images can evoke fewer negative reactions if they are interpreted as art and not as documentary (Gerger et al., 2014). Hence, from a theoretical perspective, the perception of artistic protest could mitigate or prevent reduced social identification, mobilization, public support, or engagement of policymakers, when disruption is present in protest action. In addition, while broad public support unambiguously decreases with higher levels of disruption, creative or artistic action may have positive as well as negative consequences. Overall, it seems reasonable to assume that artistic activism as a protest tactic may be capable of balancing out certain negative aspects of disruption in the context of media attention and perception, as well as public support. Together, our reflections lead us to the conclusion that artistic and disruptive tactics are less alternatives than they are combinable tactics. Considering their interplay, there ought to be a strategic equilibrium within which positive impact is possible.

4.5 An Equilibrium of Effective Disruptive Artistic Action

The aim of this chapter was to tackle the question of whether artistic protest tactics offer a strategic opportunity for social movements to push the limits of disruptive tactics. We have demonstrated that mass media play a crucial role for social movements to reach out to potential participants or supporters but also to convey their claims and demands to the public. In this relationship of social movements, mass media, and the public, artistic and disruptive tactics both have the potential to generate news media attention – whereby disruption tends to be more effective. While neither tactic appears to fully evade the protest paradigm, artistic tactics are likely to be framed less negatively than disruptive ones. Similarly, while general public support tends to decrease as normative disruption increases, artistic action – even if disruptive – may be interpreted more positively and at least not more negatively than mere disruption. Thus, these tactics are interrelated and not to be seen as dichotomous (Konya, 2021; B. J. Ryan, 2019).

The interaction between artistic and disruptive tactics relies on two premises: first, artistic actions amplify the effectiveness of disruptive actions. Second, as disruption increases, artistic tactics mitigate negativity. In this interplay, we assume a strategic equilibrium of artistic and disruptive action, in which an optimum of effectiveness – in terms of media attention, media portrayal and gaining public support – exists. On one side of the equilibrium, the degree of disruption determines tactics' effectiveness, yet it may enhance negative perceptions. On the other side, the degree to which artistic actions are recognized and appreciated as art becomes the balancing factor between tactic effectiveness and disruptiveness.

Evidently, placing art at the center of discussions about representation is never an easy task: while disruption is relatively easy to identify and even to quantify, for example, through images of blockades, numbers of arrests, casualties, or amounts of damage, classifications of artistic elements remain much vaguer. Art as a subject and, in particular, its significance for society are controversial and ambiguous issues (Belfiore & Bennett, 2007). On the one hand, many within the art world support the notion that art has an activist role in social change (Duncombe, 2016), in particular to address climate change (Buckland, 2012; McKibben, 2005). On the other hand, the art world tends to defy instrumentalization. Hence, artistic activism can also be devalued in public debates by artists or art critics (Lee, 2016; Nurmis, 2016). Even within movements, the definition of protest action as artistic can be a challenge, for example, if perceptions of the value of art differ between participating laypersons and artists (Lee & Lingo, 2011).

Nonetheless, in the relationship between movement, media, and the public sphere, the main question may not be whether creative expressions by activists are actually art. Rather, a crucial consideration is whether and to what degree they are perceived as art by journalists and the public. A public negotiation of meaning must take place where the journalist, as first observer and key intermediary between movements and the public, is able to contemplate artistic protest strategies through the intended conceptual lens of art or culture. Thus, in the design and communication of their actions, movements must ultimately strive to ensure that political journalists and subsequently the public understand symbolic, disruptive actions within a corresponding framework – as art rather than as a freak show.

The key to finding an equilibrium between disruptive and artistic tactics may ultimately reside in how much movements can effectively articulate disruptiveness in the realm of the symbolic rather than the substantive, without becoming politically irrelevant (see Fig. 4.1). In the case of peaceful protest, its political legitimacy would tend to increase as its artistic nature is recognized. At the same time, the protest's outreach would likely be limited; its relevance as a media event perhaps restricted to the arts and culture section. In contrast, protests with high degrees of disruption, violence, or crime, which would hardly be recognized as art, are

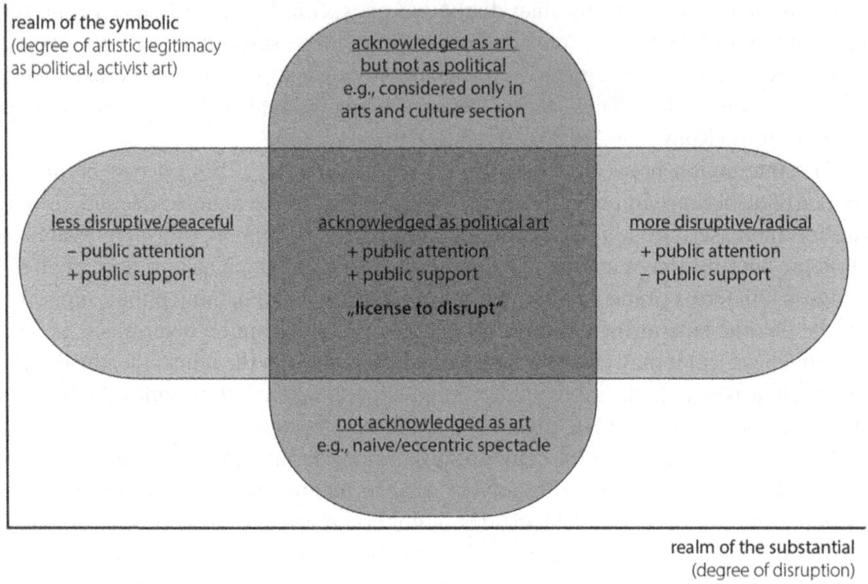

Fig. 4.1 The equilibrium of effective disruptive artistic action

extremely unlikely to be appreciated by the media and the public. Even if some artistic value was granted to such actions, movements would at best be considered as radical avant-garde, at worst as radical lunatics. Neither should be in the strategic interest of a movement. It then follows that an effective balance would lie within the limits of moderate disruption and the simultaneous recognition of political, activist art. This would allow movements to articulate dissent or defiance without ever reaching levels of crime or violence that are strongly normatively condemned. Moderate degrees of disruption are able to attract more attention if staged as an artistic performance. The more these are perceived as art, the more positively they would be processed. The symbolic nature of art and its creative relation to reality as an alterable construction may then grant artistic expressions a special license to be transgressive and disruptive while remaining largely peaceful.

However, gaining media attention at all but specially achieving legitimizing or even positive coverage remains a challenge. The consideration of artistic tactics in protests or their combination with disruptive tactics is no easy solution to desirable reporting and favorable public opinion. Mainstream media will likely default to delegitimizing frames, and despite new opportunities brought about by digital media, they still remain one of the primary channels through which social movements reach broader publics. Even so, movements can still be effective if they manage to place images that evoke positive emotions and convey a constructive message as opposed to only an oppositional one. This is not possible with images of a mere disruptive nature.

Overall, in this specific context of social movements' tactics as a mean of strategic communication, media routines, and public opinion formation, it turns out that – even if not a golden key – artistic protest tactics are a way to enhance attention and balance negative consequences of disruptive protest tactics. Naturally, these processes have to be contemplated within the larger context of protest events and also considering further movements tactics when it comes to making any strategic decisions. As an illustration, the effectiveness of disruptive artistic action not only arises within the described equilibrium but may also depend on the specific time and place, as well as the involvement of art and media experts (Shultziner & Shoshan, 2018). Furthermore, the overall visual artistic design or performative concept is relevant and can have effects on its own. It is well-established that visual content and artworks in an environmental context tend to refer to apocalyptic narratives. Although these create attention, they constrain people's willingness to engage (Nurmis, 2016; Sommer & Klöckner, 2021; Wang et al., 2018). In all, activists, artists, or artivists involved in environmental movements should carefully plan and act strategically, for example, by understanding media logics (Shoemaker & Reese, 2014; Vliegenthart & Walgrave, 2012), studying historical examples of protest action recognized as art (McGarry et al., 2020; Reed, 2019; Weibel, 2015), and referring to professional expertise in designing artistic activism (Duncombe, 2016; Duncombe & Lambert, 2017).

References

Alam, M., & Nilan, P. (2018). The campaign to save the Bandung city forest in Indonesia: A cognitive praxis analysis of protest repertoires. *Indonesia and the Malay World, 46*(136), 343–359.

Alexander, V. D., & Rueschemeyer, M. (2005). *Art and the state: The visual arts in comparative perspective*. Palgrave Macmillan.

Alkaher, I. (2020). Climate literacy and environmental activism. In W. Leal Filho, A. M. Azul, L. Brandli, P. G. Özuyar, & T. Wall (Eds.), *Climate action* (pp. 1–14). Springer International Publishing.

Ashley, L., & Olson, B. (1998). Constructing reality: Print media's framing of the women's movement, 1966 to 1986. *Journalism & Mass Communication Quarterly, 75*(2), 263–277.

Baumann, S. (2007). A general theory of artistic legitimation: How art worlds are like social movements. *Poetics, 35*(1), 47–65.

Belfiore, E., & Bennett, O. (2007). Rethinking the social impacts of the arts. *International Journal of Cultural Policy, 13*(2), 135–151.

Bishop, C. (Ed.). (2006). *Participation*. Whitechapel.

Blasch, J., & Turner, R. W. (2016). Environmental art, prior knowledge about climate change, and carbon offsets. *Journal of Environmental Studies and Sciences, 6*(4), 691–705.

Boone, G. M., Secci, J., & Gallant, L. M. (2018). Resistance: Active and creative political protest strategies. *The American Behavioral Scientist, 62*(3), 353–374.

Boyle, M. P., McLeod, D. M., & Armstrong, C. L. (2012). Adherence to the protest paradigm: The influence of protest goals and tactics on news coverage in U.S. and international newspapers. *The International Journal of Press/Politics, 17*(2), 127–144.

Branagan, M. (2003). The art(s) of nonviolent activism. *Dissent, 11*, 36–40.

Bruce, C. (2015). The balaclava as affect generator: Free pussy riot protests and transnational iconicity. *Communication and Critical/Cultural Studies, 12*(1), 42–62.

Bruner, M. L. (2005). Carnivalesque protest and the humorless state. *Text and Performance Quarterly, 25*(2), 136–155.
Buckland, D. (2012). Climate is culture. *Nature Climate Change, 11*(2), 137–140.
Bugden, D. (2020). Does climate protest work? Partisanship, protest, and sentiment pools. *Socius: Sociological Research for a Dynamic World, 6*, 1–13.
Burgess, C., & Read, R. (2020). Extinction rebellion and environmental activism – The XR interviews. *Journal of Human Rights and the Environment, 11*(3), 171–180.
Butler, J. (2015). *Notes toward a performative theory of assembly*. Harvard University Press.
Caniglia, B. S., Brulle, R. J., & Szasz, A. (2015). Civil society, social movements, and climate change. In R. E. Dunlap & R. J. Brulle (Eds.), *Climate change and society* (pp. 235–268). Oxford University Press.
Catanzaro, M., & Collin, P. (2021). Kids communicating climate change: Learning from the visual language of the SchoolStrike4Climate protests. *Educational Review*, 1–24.
Chan, J. M., & Lee, C.-C. (1984). Journalistic "paradigms" on civil protests: A case study of Hong Kong. In A. Arno & W. Dissanayake (Eds.), *The News Media in National and International Conflict* (pp. 183–202). Routledge.
Chapman, D. A., Corner, A., Webster, R., & Markowitz, E. M. (2016). Climate visuals: A mixed methods investigation of public perceptions of climate images in three countries. *Global Environmental Change, 41*, 172–182.
Cupchik, G. C. (1995). Emotion in aesthetics: Reactive and reflective models. *Poetics, 23*(1–2), 177–188.
Curtis, D. J. (2011). Towards a culture of landcare: The arts in community capacity building for natural resources management. *Journal of Environmental Assessment Policy and Management, 13*(04), 673–696.
Curtis, D. J., Mainguy, G., Reid, N., & Reeve, I. (2014). Towards ecological sustainability: Observations on the role of the arts. *Institut Veolia, 7*(1), 1–15.
Dardis, F. E. (2006). Marginalization devices in U.S. press coverage of Iraq war protest: A content analysis. *Mass Communication and Society, 9*(2), 117–135.
de Moor, J. (2017). The 'efficacy dilemma' of transnational climate activism: The case of COP21. *Environmental Politics, 27*(6), 1079–1100.
Demos, T. (2016). Between rebel creativity and reification: For and against visual activism. *Journal of Visual Culture, 15*(1), 85–102.
Doerr, N., Mattoni, A., & Teune, S. (2013). Toward a visual analysis of social movements, conflict, and political mobilization. *Research in Social Movements, Conflicts and Change, 35*, xi–xxvi.
Duncombe, S. (2016). Does it work? The æffect of activist art. *Social Research: An International Quarterly, 83*(1), 115–134.
Duncombe, S., & Lambert, S. (2017). Lessons from utopia. *Visual Inquiry, 6*(2), 253–272.
Entman, R. M. (1993). Framing: Toward clarification of a fractured paradigm. *Journal of Communication, 43*(4), 51–58.
Feinberg, M., Willer, R., & Kovacheff, C. (2020). The activist's dilemma: Extreme protest actions reduce popular support for social movements. *Journal of Personality and Social Psychology, 119*(5), 1086–1111.
Fichnová, K., Wojciechowski, Ł. P., & Szabo, P. (2019). A reflection on the citizens' attitudes to the political situation (and their creative renditions). *Communication Today, 10*(1), 76–93.
Foderaro, L. W. (2014, September 21). Taking a call for climate change to the streets. *The New York Times*. Retrieved from http://www.nytimes.com/2014/09/22/nyregion/new-york-city-climatechange-march.html?_r=0
Fuentes, M. A. (2012). 'Investments towards returns': Protest and performance in the era of financial crises. *Journal of Latin American Cultural Studies, 21*(3), 449–468.
Gamson, W. A. (2004). Bystanders, public opinion, and the media. *The Blackwell Companion to Social Movements*, 242–261.

Geise, S., Heck, A., & Panke, D. (2021). The effects of digital media images on political participation online: Results of an eye-tracking experiment integrating individual perceptions of "photo news factors". *Policy and Internet, 13*(1), 54–85.

Gerger, G., Leder, H., & Kremer, A. (2014). Context effects on emotional and aesthetic evaluations of artworks and IAPS pictures. *Acta Psychologica, 151*, 174–183.

Gil-Lopez, T. (2021). Mainstream protest reporting in the contemporary media environment: Exploring (in) stability and adherence to protest paradigm from 1998 to 2017. *Journalism & Mass Communication Quarterly, 98*(3), 692–724.

Gitlin, T. (1980). *The whole world is watching: Mass media in the making and unmaking of the new left*. University of California Press.

Grindon, G. (2011). Surrealism, dada, and the refusal of work: Autonomy, activism, and social participation in the radical Avant-Garde. *Oxford Art Journal, 34*(1), 79–96.

Gutting, R. S. (2019). Contentious activities, disrespectful protesters: Effect of protest context on protest support and mobilization across ideology and authoritarianism. Political Behaviour, 1–26.

Harcup, T., & O'Neill, D. (2001). What is news? Galtung and Ruge revisited. *Journalism Studies, 2*(2), 261–280.

Harlow, S., & Johnson, T. J. (2011). The Arab springl overthrowing the protest paradigm? How the New York Times, global voices and twitter covered the Egyptian revolution. *International Journal of Communication, 5*, 16.

Harlow, S., Salaverría, R., Kilgo, D. K., & García-Perdomo, V. (2017). Protest paradigm in multimedia: Social media sharing of coverage about the crime of Ayotzinapa, Mexico. *Journal of Communication, 67*(3), 328–349.

Harold, C. (2004). Pranking rhetoric: "Culture jamming" as media activism. *Critical Studies in Media Communication, 21*(3), 189–211.

Hsiao, Y., & Radnitz, S. (2020). Allies or agitators? How partisan identity shapes public opinion about violent or nonviolent protests. *Political Communication*, 1–19.

Imposti, G. E. (2015). Dissident "street art" resisting – Neo-soviet discourse: The "Voina" and "Pussy Riot" groups! *Street Art, 5*(10), 1–26.

Kananovich, V. (2016). Progressive artists, political martyrs, or blasphemous hussies? A content analysis of the Russian media coverage of the pussy riot affair. *Popular Music and Society, 39*(4), 396–409.

Kay, A. (2000). Art and community development: The role the arts have in regenerating communities. *Community Development Journal, 35*(4), 414–424.

Kenix, L. J. (2008). Conscientious dissident or radical eccentric: Negotiation of mediated protest images in New Zealand. *MEDIANZ: Media Studies Journal of Aotearoa New Zealand, 11*(2), 1–32.

Kilgo, D. K., Harlow, S., García-Perdomo, V., & Salaverría, R. (2018). From #Ferguson to #Ayotzinapa: Analysing differences in domestic and foreign protest news shared on social media. *Mass Communication and Society, 21*(5), 606–630.

Kleut, J., & Milojevic, A. (2021). Framing protest in online news and readers' comments: The case of Serbian protest "against dictatorship". *International Journal of Communication, 15*, 21.

Konya, N. (2021). Breaking billboards: Protest and a politics of play. *Contemporary Political Theory, 20*(2), 250–271.

Krause, A., & Bucy, E. P. (2018). Interpreting images of fracking: How visual frames and standing attitudes shape perceptions of environmental risk and economic benefit. *Environmental Communication, 12*(3), 322–343.

Lavender, A. (2019). Theatricalizing protest: The chorus of the commons. *Performance Research, 24*(8), 4–11.

Lee, C. W. (2016). The arts and crafts of participatory reforms: How can socially engaged art and public deliberation inform each other? *FIELD, 3*(Spring), 7–37.

Lee, F. L. (2014). Talk radio, the mainstream press, and public opinion in Hong Kong. Hong Kong University Press.

Lee, C. W., & Lingo, E. L. (2011). The "got art?" paradox: Questioning the value of art in collective action. *Poetics, 39*(4), 316–335.

McFarlane, T., & Hay, I. (2003). The battle for Seattle: Protest and popular geopolitics in The Australian newspaper. *Political Geography, 22*(2), 211–232.

McGarry, A., Erhart, I., Eslen-Ziya, H., Jenzen, O., & Korkut, U. (2020). Introduction: The aesthetics of global protest: Visual culture and communication. In A. McGarry, I. Erhart, E.-Z. Hande, O. Jenzen, & U. Korkut (Eds.), *The aesthetics of global protest: Visual culture and communication* (pp. 15–35). Amsterdam University Press.

McKibben, B. (2005). Imagine that: What the warming world needs now is art sweet art. Grist.Org. www.grist.org/article/mckibben-imagine

McLeod, D. M. (1995). Communicating deviance: The effects of television news coverage of social protest. *Journal of Broadcasting & Electronic Media, 39*(1), 4–19.

McLeod, D. M. (2007). News coverage and social protest: How the media's protect paradigm exacerbates social conflict. *Journal of Dispute Resolution, 2007*(1), 185–194. https://scholarship.law.missouri.edu/jdr/vol2007/iss1/12

McLeod, D. M., & Detenber, B. H. (1999). Framing effects of television news coverage of social protest. *Journal of Communication, 49*(3), 3–23.

McLeod, D. M., & Hertog, J. K. (1999). Social control, social change and the mass media's role in the regulation of protest groups. *Mass Media, Social Control, and Social Change: A Macrosocial Perspective*, 305–330.

Metag, J., Schäfer, M. S., Füchslin, T., Barsuhn, T., & Kleinen-von Königslöw, K. (2016). Perceptions of climate change imagery: Evoked salience and self-efficacy in Germany, Switzerland, and Austria. *Science Communication, 38*(2), 197–227.

Myers, D. J., & Caniglia, B. S. (2004). All the rioting that's fit to print: Selection effects in national newspaper coverage of civil disorders, 1968–1969. *American Sociological Review, 69*(4), 519–543.

Nash, S. L., & Steurer, R. (2021). Climate change acts in Scotland, Austria, Denmark and Sweden: The role of discourse and deliberation. *Climate Policy*, 1–12.

Neumayer, C., & Rossi, L. (2018). Images of protest in social media: Struggle over visibility and visual narratives. *New Media & Society, 20*(11), 4293–4310.

NicGhabhann, N. (2018). City walls, bathroom stalls and tweeting the Taoiseach: The aesthetics of protest and the campaign for abortion rights in the Republic of Ireland. *Continuum, 32*(5), 553–568.

Nurmis, J. (2016). Visual climate change art 2005-2015: Discourse and practice. *Wiley Interdisciplinary Reviews: Climate Change, 7*(4), 501–516.

O'Neill, K. (2004). Transnational protest: States, circuses, and conflict at the frontline of global politics. *International Studies Review, 6*(2), 233–251.

O'Neill, S. J., Boykoff, M., Niemeyer, S., & Day, S. A. (2013). On the use of imagery for climate change engagement. *Global Environmental Change, 23*(2), 413–421.

Orazani, S. N., & Leidner, B. (2019). The power of nonviolence: Confirming and explaining the success of nonviolent (rather than violent) political movements. *European Journal of Social Psychology, 49*(4), 688–704.

Otto, I. M., Donges, J. F., Cremades, R., Bhowmik, A., Hewitt, R. J., Lucht, W., Rockström, J., Allerberger, F., McCaffrey, M., Doe, S. S. P., Lenferna, A., Morán, N., van Vuuren, D. P., & Schellnhuber, H. J. (2020). Social tipping dynamics for stabilizing Earth's climate by 2050. *Proceedings of the National Academy of Sciences, 117*(5), 2354–2365.

Pelowski, M., Markey, P. S., Forster, M., Gerger, G., & Leder, H. (2017). Move me, astonish me… delight my eyes and brain: The Vienna Integrated Model of top-down and bottom-up processes in Art Perception (VIMAP) and corresponding affective, evaluative, and neurophysiological correlates. *Physics of Life Reviews, 21*, 80–125.

PytlikZillig, L. M., Hutchens, M. J., Muhlberger, P., Gonzalez, F. J., & Tomkins, A. J. (2018). *Deliberative public engagement with science*. Springer International Publishing.

Reed, T. V. (2019). *The art of protest: Culture and activism from the Civil Rights Movement to the present*. University of Minnesota Press.

Robertson, K. (2006). Chapter 16. Taking Québec City: Protest, carnival and tourism at the summit of the Americas. In D. Picard & M. Robinson (Eds.), *Festivals, tourism and social change: Remaking worlds* (pp. 269–283). Blue Ridge Summit: Channel View Publications.

Rössler, P., Bomhoff, J., Haschke, J. F., Kersten, J., & Müller, R. (2011). Selection and impact of press photography. An empirical study on the basis of photo news factors. *Communications, 36*(4), 415–439.

Ryan, H. E. (2015). Affect's effects: Considering art-activism and the 2001 crisis in Argentina. *Social Movement Studies, 14*(1), 42–57.

Ryan, B. J. (2019). Standing acts: The political aesthetics of defiant resistance. *International Political Sociology, 13*(2), 111–127.

Sanz, T., & Rodriguez-Labajos, B. (2021). Does artistic activism change anything? Strategic and transformative effects of arts in anti-coal struggles in Oakland, CA. *Geoforum, 122*, 41–54.

Shoemaker, P. J. (1982). The perceived legitimacy of deviant political groups: Two experiments on media effects. *Communication Research, 9*(2), 249–286.

Shoemaker, P. J. (1984). Media treatment of deviant political groups. *Journalism Quarterly, 61*(1), 66–82.

Shoemaker, P. J., & Reese, S. D. (2014). *Mediating the message in the 21st century: A media sociology perspective* (3rd ed.). Routledge/Taylor & Francis Group.

Shultziner, D., & Shoshan, A. (2018). A journalists' protest? Personal identification and journalistic activism in the Israel social justice protest movement. *The International Journal of Press/Politics, 23*(1), 44–69.

Simon, B., & Klandermans, B. (2001). Politicized collective identity: A social psychological analysis. *American Psychologist, 56*(4), 319–331.

Simpson, B., Willer, R., & Feinberg, M. (2018). Does violent protest backfire? Testing a theory of public reactions to activist violence. *Socius, 4*, 2378023118803189.

Smith, J., McCarthy, J. D., McPhail, C., & Augustyn, B. (2001). From protest to agenda building: Description bias in media coverage of protest events in Washington, DC. *Social Forces, 79*(4), 1397–1423.

Sommer, L. K., & Klöckner, C. A. (2021). Does activist art have the capacity to raise awareness in audiences?—A study on climate change art at the ArtCOP21 event in Paris. *Psychology of Aesthetics, Creativity, and the Arts, 15*(1), 60–75.

Stanisavljevic, M. (2016). Protest communication between mass media and aesthetics. *Österreichische Zeitschrift Für Soziologie, 41*(2), 123–148.

Steinholt, Y. B. (2013). Kitten heresy: Lost contexts of pussy Riot's punk prayer. *Popular Music and Society, 36*(1), 120–124.

Stürmer, S., & Simon, B. (2004). Collective action: Towards a dual-pathway model. *European Review of Social Psychology, 15*(1), 59–99.

Tausch, N., Becker, J. C., Spears, R., Christ, O., Saab, R., Singh, P., & Siddiqui, R. N. (2011). Explaining radical group behaviour: Developing emotion and efficacy routes to normative and nonnormative collective action. *Journal of Personality and Social Psychology, 101*(1), 129–148.

Taylor, V., & Van Dyke, N. (2004). "'Get up, stand up'": Tactical repertoires of social movements. In D. A. Snow, S. A. Soule, & H. Kriesi (Eds.), *The Blackwell companion to social movements* (pp. 262–293). Blackwell Publishing.

Teune, S. (2005, September 8–10). *Art and the re-invention of political protest*. Paper presented at the 3rd ECPR conference. Budapest. https://ecpr.eu/Events/Event/PaperDetails/19594

Thomas, E. F., & Louis, W. R. (2014). When will collective action be effective? Violent and non-violent protests differentially influence perceptions of legitimacy and efficacy among sympathizers. *Personality and Social Psychology Bulletin, 40*(2), 263–276.

Trifonov, S. V. (2017). Twenty-five years of democracy, twenty-five years of social protest: The role of the carnivalesque in Bulgaria's 2013 antigovernment protests. *Journal of International and Intercultural Communication, 10*(3), 237–254.

Tunalı, T. (2020). Humour as political aesthetics in street protests during the political Ice Age. *The European Journal of Humour Research, 8*(2), 129–145.

van Zomeren, M., Postmes, T., & Spears, R. (2008). Toward an integrative social identity model of collective action: A quantitative research synthesis of three socio-psychological perspectives. *Psychological Bulletin, 134*(4), 504–535.

Veneti, A. (2017). Aesthetics of protest: An examination of the photojournalistic approach to protest imagery. *Visual Communication, 16*(3), 279–298.

Vliegenthart, R., & Walgrave, S. (2012). The interdependency of mass media and social movements. In H. Semetko & M. Scammell (Eds.), *The SAGE handbook of political communication* (pp. 387–398). SAGE Publications Ltd.

von Zabern, L., & Tulloch, C. D. (2021). Rebel with a cause: The framing of climate change and intergenerational justice in the German press treatment of the Fridays for future protests. *Media, Culture & Society, 43*(1), 23–47.

Wang, S., Corner, A., Chapman, D., & Markowitz, E. (2018). Public engagement with climate imagery in a changing digital landscape. *Wiley Interdisciplinary Reviews: Climate Change, 9*(2), e509.

Weibel, P. (2015). People, politics, and power. In P. Weibel (Ed.), *Global activism: Art and conflict in the 21st century*. [exhibition, Karlsruhe, ZKM Centre for Art and Media, December 14, 2013—March 30, 2014] (pp. 29–61). The MIT Press.

Weij, F., Berkers, P., & Engelbert, J. (2015). Western solidarity with Pussy Riot and the Twittering of cosmopolitan selves: Western solidarity with Pussy Riot. *International Journal of Consumer Studies, 39*(5), 489–494.

Wouters, R. (2013). From the street to the screen: Characteristics of protest events as determinants of television news coverage. *Mobilization: An International Quarterly, 18*(1), 83–105.

Wouters, R. (2019). The persuasive power of protest. How protest wins public support. *Social Forces, 98*(1), 403–426.

Xu, K. (2013). Framing Occupy Wall Street: A content analysis of the New York Times and USA Today. *International Journal of Communication, 7*, 2412–2432.

Chapter 5
Can We Be Entertained to Change Our Lives?: An Introduction to Games for Increasing Environmental Awareness

Kristoffer S. Fjællingsdal and Christian A. Klöckner

5.1 Introduction

In April 2007, there was a massive oil shortage in the United States, heavily affecting both the access to gas as well as bringing major companies to the brink of bankruptcy due to severely heightened fuel costs and diminished operating capacity. A committee was soon established to find creative solutions to the shortage, consisting of approximately 2000 individuals from all 50 US states and diverse walks of life – ranging from hobby farmers and soldiers to art students and employees at General Motors. In addition to the major societal issues arising with a shortage of oil, this committee also had to consider how everyday behaviors such as getting to work, preparing dinner, or even socializing with friends and family should be conducted. For each day of the 6-week-long crisis, new issues arose – airlines would cancel their flights, food shortages arose due to delivery vans not being able to restock produce, and public transportation broke down under the new pressure from people who could no longer afford using their own cars. The committee would eventually release an online document containing their proposed solutions to some of the problems the oil shortage had brought along and included such examples as how the architectural industry, neighborhoods, parenting, and motorsports could make the transition into a shift toward a more sustainable future.

The "kicker" here is that none of this happened in real life[1] – it was a massive online alternate reality game (or ARG) called *World Without Oil* (or WWO), invented and conducted by writers, game designers, and laypeople with funding from the Corporation for Public Broadcasting (McGonigal, 2012, p. 303). Observed through the lens of disruptive communication, the individuals and groups involved in the WWO project actively went out of their way to think about alternative solutions and pathways to circumventing the oil crisis that they were faced with in the

[1] Frighteningly, much of that scenario does not sound so unrealistic in summer 2022.

C. A. Klöckner, E. Löfström, *Disruptive Environmental Communication*, Psychology and Our Planet, https://doi.org/10.1007/978-3-031-17165-9_5

digital world, thus certainly confronting the status quo or *business as usual* aspect of their everyday lives. They would furthermore, in various ways, *confront the commonly accepted assumption* that every aspect of human civilization was entirely dependent on oil to function. Lastly, the *emotional involvement* in the project was clearly palpable – especially considering that so many people actively and voluntarily came together to brainstorm a great deal of innovative ideas and solutions for an oil-free world. Going back to the formal definition of disruptive communication presented in the first chapter of this book, it soon becomes clear that games – in their various sizes, shapes, and formats – can be used in disruptive communication interventions.

Today, however, despite its initial success and media coverage, the WWO project is seldomly spoken of, and the empirical foundation on the efficiency of environmental games in sustainability education is lacking (Hallinger et al., 2020). The same can be said for other game-based learning interventions as well, and media narratives often highlight research that identifies them as potential sources of aggression (e.g., Bushman & Anderson, 2001) or addiction (e.g., van Rooij et al., 2010). Despite this, however, the interest in developing games with an environmental theme or area of focus has gradually increased among academics, game designers, and nongovernmental organizations alike (Reckien & Eisenack, 2013), particularly games that explicitly wish to do something more than just entertain their players. On the environmental side of such educational games, traditionally known as *serious games* (Abt, 1970), various sophisticated examples exist in both digital and physical formats. Commercial environmental video games such as *Fate of the World* (Roberts, 2011) and *Eco* (Strange Loop Games, 2020) put the players in charge of maintaining balance in digital climate politics and ecosystem management, respectively, while board games such as the Oil Springs scenario of Settlers of Catan (Assadourian & Hansen, 2011) and Keep Cool (Eisenack & Petschel-Held, 2004) revolve around subjects such as the use and abuse of fossil fuels and climate negotiations. Other environmental games deal with sustainability issues such as water saving, waste management, and natural disasters on a very general level (Ouariachi et al., 2018) or feature more specific topics such as the ecological threat of invasive seaweed (Skukan et al., 2020).

The variety of environmental issues is considerable, but what may be the potential benefits and weaknesses of using games to learn about such complex topics? The next sections of this chapter will explore some of these questions in depth. The chapter will then conclude with preliminary suggested guidelines for educators who wish to use environmental games in disruptive communication interventions.

5.2 The Strengths and Benefits of Gaming the Environment

Before moving on with this chapter, try thinking about how many pro-environmental adverts, campaigns, pamphlets, brochures, and even commercials you have seen, heard, or otherwise noticed in the past 6 months. Now ask yourself: how many of these do I remember clearly? Many of you may find that you have trouble remembering or even grasping the material that the advert or campaign wished to teach you in the first place. Within the field of environmental communication, this is nothing new. All too often, environmental information campaigns, while occasionally informative and well-meaning, generally tend to be too unspecific or lacking in terms of what they seek to communicate (Klöckner, 2015, p. 165). Despite this, they can be found just about anywhere – bus stops, on the television, as pop-ups or ads on the Internet – as well as in a variety of different formats: popular science books, documentaries, and scientific charts, just to name a few. When faced with such an overwhelming number of communications, the human brain tends to start blocking out the intended message, as well-meaning as it may be – a term known as *environmental numbness* (Gifford, 2011). Most information campaigns, while initially perceived as shocking and disturbing or even hopeful and optimistic, will eventually lose their initial appeal and become static, droll statements running on repeat across a variety of different platforms, an effect also seen in massive health campaigns (e.g., Smith et al., 2002). The result, ultimately, is that people become highly knowledgeable about environmental issues, yet little to no pro-environmental action is taken. Furthermore, as much as they may hear or read about them, many people have little to no direct experience with environmental issues and their visible effects (Myers et al., 2013), instead learning about them through noninteractive media such as documentaries and films (e.g., Greitemeyer, 2013), newspapers (Reis, 1999), religious or politically affiliated outlets (e.g., Greeley, 1993; McCright et al., 2014), or tiny portions of selected school curriculums (Huckle, 1993). As such, a psychological distance is formed – climate change and its negative effects is something that is happening "over there," not "here" (Stoknes, 2017, p. 123).

Although environmental adverts are represented in anything from newspapers (e.g., Dimopoulos & Koulaidis, 2003) to radio broadcasts (Klöckner, 2015, p. 121), they are seldomly *memorable* – meaning they do not last long when faced with the complexities of the human memory. By contrast, games are designed to provide lasting experiences (Schell, 2008) and are often played for hours on end across the entire globe (McGonigal, 2012). As such, the interest in using games for scientific communication becomes obvious: dedicating time, effort, and attention to social issues through entertaining gameplay experiences should, at least in theory, be more effective at inspiring and motivating people to act pro-environmentally than static forms of media that are only viewed for mere seconds or minutes. The attention-grabbing and memorable nature of games certainly has an empirical foundation, primarily due to their inherent immersive qualities. When *immersed* into a game, players report that they often lose the track of time and "get sucked into" the game world (Brown & Cairns, 2004), thus ensuring that their attention is directed to its fullest extent on what is going on in the game – as well as what it is designed to teach.

Immersion is, in fact, a core motivational factor as to why people choose to play games to begin with (Jennett et al., 2008; Sweetser & Wyeth, 2005) and generally refers to a temporary suspended disregard for certain aspects of "real life" such as physical surroundings, flow of time, and awareness of the physical self (Brown &

Cairns, 2004; Sweetser & Wyeth, 2005) in favor of what goes on in the game world (i.e., the screen or the board). As a result of experiencing game-based immersion, the player gets a sense of presence, agency, and impact on the game world, or a feeling that they are somewhat "there" in the game itself. Although commonly discussed as a holistic phenomenon in most research literature, however, immersion is arguably a multifaceted psychological construct consisting of at least three primary forms: sensory, challenge-based, and imaginative immersion (Ermi & Mäyrä, 2005). *Sensory immersion* refers to the game's audiovisual elements that engross the player and enable their journey into the game world, *challenge-based immersion* occurs through the obstacles and puzzles in the game that the player must overcome, and *imaginative immersion* happens when a player can use their imagination, empathize, and gain emotional involvement in the characters and story or plainly enjoy the fantasy aspects of the game world (Ermi & Mäyrä, 2005). Immersion is furthermore theorized to be stage-based, leading the player from basic engagement (learning the game's controls and gaining appreciation of the game's theme) through engrossment (emotional involvement in the game) and eventually into total immersion or presence (psychological detachment from physical reality and attentional shift to the game world (Brown & Cairns, 2004)). As this immersion created by (good) games interferes with people's lives outside games, the game experience has the potential to be disruptive in the sense that it takes people out of their everyday lives.

Immersion furthermore bears a strong resemblance to another psychological factor that is frequently discussed in game design – a phenomenon known simply as flow. The term *flow phenomenon* (or *flow state*) is commonly attributed to the works of the Hungarian-American psychologist Mihaly Csikszentmihalyi and can simply be described as complete absorption into an activity with clearly defined goals, feedback, and an optimal balance between the difficulty level of the task and the task performer's skill level (Nakamura & Csikszentmihalyi, 2014). Csikszentmihalyi observed the creative processes of numerous individuals conducting various interest-based activities in the 1960s and frequently noticed that they tended to suppress physical needs such as nourishment and rest when the activity was perceived as engaging or engrossing. This bears some resemblance to descriptions of immersion, as noted in the previous paragraph – often to the point where some researchers have described flow as a *form* of immersion (e.g., Jennett et al., 2008; Weibel & Wissmath, 2011). Flow, however, is sometimes described as complete involvement into an activity, whereas immersion generally refers to a loss of (or cognitive disconnect from) a physical context in favor of a virtual one (Nacke & Lindley, 2008) – often *without* a *complete* disregard of one's physical environment (Jennett et al., 2008). Regardless of the rather vague definition in research literature, immersion (and, by extension, something akin to a flow state) remains a very important component of enjoyable gameplay experiences – regardless of the game's topic.

While immersion is a key concern in the field of game design, the *invisibility of environmental decline* (Hansen & Machin, 2013) and *indirect rather than direct experiences with environmental issues* (Fazio & Zanna, 1978) are two key concerns within the field of environmental communication. This is where games can play important roles, both by portraying something that is literally invisible to the naked eye, such as CO_2 (Sheppard, 2012, p. 3) as well as providing simulated, immersive, and (at least somewhat) realistic experiences of how one's personal impact on the environment can transpire. Games (both digital and board-based) are known to take complex topics and systems, such as global warming and ecosystem balance, and simplify them to make their players gain a better understanding of how their real-life equivalents work (e.g., Fjællingsdal & Klöckner, 2019, 2020). The way in which they do so is by essentially "compressing" what would otherwise be a rather unmanageable load of information into a clearer and more lucid, albeit simplistic version of the subject or topic in question – a so-called microworld (Egenfeldt-Nielsen et al., 2013, p. 237). The game experience then turns this microworld into the center of the players attention, disrupting their established and complex "outside world," where the environmental issues go unattended. Elements that would likely make the topic too complex to understand for a layperson, as well as the scientific lingo and jargon that constitute core components of scientific research articles, are normally omitted in favor of colorful and lush graphics, deep and emotional narratives, a visualization of one's personal impact on the virtual environment, and so on. By capturing the players' attention and giving them a sense of agency or importance to how the virtual world develops, games can therefore be used to increase what psychologists normally refer to as *perceived behavioral control* (or PBC; Ajzen, 1991) – a person's set of beliefs, resources, opportunities, and abilities to perform a specific (pro-environmental) behavior (Klöckner, 2015, p. 70). Ideally, although perhaps merely as a singular joint in a series of interventions, games could therefore inspire their players to act in the real world based on their experiences in the virtual world (Peters et al., 1998). A game designer's freedom to create narratives, experiences, or visualizations can also be used to create disruptive events (e.g., a suddenly appearing opponent, a new challenge that changes the rules, or a meteor approaching) even within the game's microworld, a technique that is often used to drive the game's narrative forward. Seen like this, games can not only be disruptive elements themselves by transporting players out of their everyday lives into a state of reflection but also use disruptive elements within their own reference framework, which is often more flexible than in the "real world."

Another strong argument for using games to educate about environment and sustainability is their inherent *multimodality*, meaning that they appeal to several of our senses and functionalities with various types of information and sensory inputs (Burn, 2008). Unlike books, which normally feature readable material in combination with static visual imagery, games can also apply animations, music, interactivity, cleverly designed puzzles and levels, or even the ability to use bodily movement for their players to learn about a given subject. Games can therefore constitute much more dynamic and self-regulated forms of learning than more traditional forms of media that are used in educational contexts. Furthermore, games are not bound by the physical or societal rules of reality and can create experiences which go beyond what could be experienced in real life. In and by itself, this does not mean that games are necessarily superior to traditional media but rather might serve a complementary function by illustrating, visualizing, and enabling the enactment of that, which books and documentaries cannot – such as by having players experiment and enact what might happen if they deforest an entire ecosystem in Eco (Strange Loop Games, 2020). The interactive and experiential aspects in relation to the virtual setting are of course crucial: while deforestation in the game will not impact the player in real life, it is still capable of portraying what might feasibly happen were deforestation to happen in the real world. Games allow their players to experience dangerous, inaccessible, or even catastrophic situations in the safe confines of virtual reality (e.g., García-Barros et al., 2015; McGonigal, 2012, p. 303), thus allowing the players themselves (possibly and preferably together with a skilled facilitator or teacher) to draw connections and implications from the game to reality.

Returning to the primary concept introduced in this book, disruptive communication, it is obvious that games, in their many formats, are voluntary and pleasurable rather than outright disruptive activities (McGonigal, 2012, p. 21). However, that alone does not change the fact that games can generate new knowledge or reinforce our knowledge about how the environment works (Fjællingsdal & Klöckner, 2019, 2020), make us question the status quo and envision alternative futures (Wu & Lee, 2015), and provoke deep thoughts about the relationship between humans and nature (Fjællingsdal & Klöckner, 2017). They can furthermore confront and frustrate us by showing clear, visual depictions between human activity, increased carbon emissions, and the subsequent negative consequences of living "business as usual" in a world where business is becoming anything but "usual." Being entertained and frustrated by something that is trying to convince or persuade you to act in a certain way might seem strange, but frustration and other negative emotions are commonly encountered by media researchers as some of the most pertinent reasons as to why people choose to consume certain types of media (Bartsch & Viehoff, 2010) – games included (Salen & Zimmerman, 2004, p. 334). Frustration and the disruption of everyday life are simply taken for granted in the world of games, perhaps more so than anywhere else in the sphere of human existence. As such, games certainly represent intriguing tools for disruptive communication practices – even going so far as to add the element of *voluntary participation* to the mix (McGonigal, 2012, p. 21).

5.3 The Weaknesses and Barriers to Gaming the Environment

Principally, the use of games in sustainability education might seem illogical. Part of an article headline from the limited empirical research that exists on environmental games asks the pertinent question "should we play games where energy is concerned?" (Boomsma et al., 2018), and, extending this question further, one might ask if we should be playing and toying with what is commonly described as some of the biggest, most complicated threats to the planet of all time – that is, environmental issues in general. Indeed, although games certainly come with a myriad of strengths and benefits in environmental communication practices (see Sect. 5.2), they also have weaknesses and barriers that need to be considered if a game-based intervention is to be successful. One of the more obvious barriers to using games to teach about the environment is that they are often costly to produce. The requirements for making an average video game, for instance, involves large-scale interdisciplinary collaboration between a wide variety of professions – ranging from 3D graphics and server optimization to storytelling and community management (Yee, 2014, p. 215) – each of which comes with an individual pay check. Board games, by comparison, often require less manpower to make but in return often consist of physical components that can, on occasion, be expensive. To cut such costs, many environmental games tend to be developed largely by well-meaning academic institutions or nongovernmental organizations (NGOs; Reckien & Eisenack, 2013) using whichever materials that are at hand, and often without much insight from professional game designers. While it can certainly cut expenses, this strategy fails to consider the importance of professional game designers and their ability to include elements that make the game enjoyable to play. Overcoming the barrier of high costs therefore often depends on successful crowdfunding, as in the case of Eco (Strange Loop Games, 2020) or the board game "CO_2" (Lacerda, 2012), or thinking creatively about how the game design process should be conducted – such as by utilizing open-source technology for making a video game (Yee, 2014) or by modifying existing games to include an environmental component (e.g., Assadourian & Hansen, 2011; Illingworth, 2020).

In addition to being costly, games are also "just games." Regardless of what one might do in the virtual world or on the tabletop, it of course does not immediately translate itself into a real impact on the real world (Fjællingsdal & Klöckner, 2020). As such, environmental games stand the risk of becoming a form of *slacktivism* – the feeling of having done something impactful merely by playing rather than the game inspiring its players to get involved and act in the real world (Sandbrook et al., 2015). Furthermore, there is little to no guarantee that the role one plays in a game can be played in a real-life setting. Where game characters are often extremely empowered and exist beyond the boundaries of what the real world allows, the players themselves often have comparatively little agency regarding how, when, and to what degree they can engage in pro-environmental behavior in a real-life setting. This discrepancy between player and avatar is part of the *simulation gap* (Bogost, 2010, p. 43) and represents one of the key challenges in employing successful game-based learning. To circumvent this simulation gap, it is important that games in some way, shape, or form inspire the players to take what they learn in the game and apply it to a real-life setting (Peters et al., 1998). Expecting every single player of a sustainability-oriented game to immediately and uncritically absorb the game's message and translate it into real-life action in such a manner, however, is likely little more than wishful thinking. Despite this, it is worth noting that the sphere in which game-based learning takes place is quite porous in that different players perceive and integrate the game's message differently (Schell, 2008, p. 10). Factors that are likely to impact the learning outcomes of playing environmental games are therefore highly varied, ranging from the players' demographic variables (age, gender, sociocultural background), previous experience and overall preference for specific game types, motivation to play, and preexisting knowledge, attitudes, values, beliefs, and norms surrounding environmental topics (Fjællingsdal & Klöckner, 2017).

The aspects of the player are certainly important in determining the efficiency of game-based learning but so are the inherent qualities and design of the game. As previously mentioned, until recently, most environmental games were developed by academics and NGOs (Reckien & Eisenack, 2013) – institutions that, by and large, do not necessarily prioritize the importance of proper game design principles. The educational foundation might be there, but the entertaining elements that constitute the core of a pleasurable gameplay experience are often lacking (Klöckner, 2015). Games that are perceived as boring, low-quality, or overly frustrating are often simply put down or not played at all (Sweetser & Wyeth, 2005). Among the critiques directed toward serious games are their high simplicity and short playtimes (Koehler et al., 2016), low production levels and graphical quality (Rice, 2006), and their use as smokescreens for conventional didactic methods – i.e., the equivalent of "chocolate-covered broccoli" (Galarneau, 2005). These trends are likely to change, however, as serious games are increasingly being co-developed in interdisciplinary teams of academics (providing the scientific foundation behind the game) and game designers (providing fun and entertaining game mechanics).

Another possible "weakness" of using environmental games in teaching is their overreliance on increasing their players' knowledge about environmental issues. This does not mean that environmental knowledge is unimportant – in fact, it is quite the opposite. Knowledge is a core component and driving force of most of the psychological frameworks that are used today to understand how pro-environmental behavior is initiated (Klöckner, 2015). However, simply knowing that a problem exists is, overall, seldomly enough to initiate pro-environmental action and behavior (Hines et al., 1987; Kollmuss & Agyeman, 2002; Moser, 2010). Despite this notion, most serious environmental games are aimed primarily toward precisely that – increasing their players' knowledge (Klöckner, 2015, p. 203; Reckien & Eisenack, 2013). In so doing, they risk excluding other important psychological variables that are important in causing pro-environmental behavior – ranging from personal attitudes, norms, emotions, and values to beliefs and habits (see Klöckner, 2015, for an extensive review). To be effective in generating pro-environmental behavior, it is therefore arguable that future environmental games need to target these psychological variables as well (Ouariachi et al., 2018).

Lastly, it is important to acknowledge that implementing games as tools for environmental communication often simply is not feasible. In a school setting, for example, existing games can be difficult to integrate with the planned curriculum of the students or pupils (e.g., Fjællingsdal & Klöckner, 2019), there might be rules and regulations pertaining to potential security issues when opening the school network for online serious gaming (Skaug et al., 2020), or the games and the required equipment to play them might simply be too costly to purchase. Also, although games are played by millions worldwide (McGonigal, 2012), there is still public hesitation surrounding the use of games in sustainability education due to their relative "newness" in the media landscape. As such, large-scale serious game interventions featuring environmental themes and topics akin to that of WWO (McGonigal, 2012, p. 303) have yet to become commonplace.

5.4 Evaluating Games in Environmental Education

So far in this chapter, we have presented research-based core strengths and weaknesses to using games in the promotion of environmental literacy. While these lists are hardly exhaustive, they should provide a general insight into how games may or may not be used to educate the public about the environment and may serve as crystalizing cores for disruptive experiences. Conclusively, it is possible to suggest that the potential use of games for such purposes is highly contextual. In discussing the relation between the player and the game, game designer Jesse Schell (2008) notes that all player experiences are inherently unique. Even if a game is met with either universal acclaim or disdain, there is still a huge variety of different ways in which its players perceive it as being either good, bad, or a combination of both. In the same vein, as people have different strategies for obtaining and integrating information, one might also expect that any single serious environmental game can lead to a series of unique individual learning outcomes that, depending on the educator or researcher studying them, may or may not be considered when the game is eventually evaluated. As an example, let us consider the game Eco, by Strange Loop Games (2020). A hypothetical way of using Eco is to distribute it to a class of 30 pupils, with the explicit instruction that they need to meet a series of in-game goals. Once the pupils have done what they are told and the game is evaluated with a thorough debriefing session between the teacher and the pupils themselves, one might find that only 6 of the 30 pupils managed to meet the goals that were set for them. If one were to judge the game's educational effectiveness simply on these measurements, it would be tempting to conclude that the game is rather inefficient, and it would likely get scrapped in the future in favor of more traditional learning methods.

However, is it safe to conclude that the remaining 24 pupils learned nothing from playing the game? As mentioned in Sect. 5.2, games promote and enable self-regulated learning, where the learners are encouraged to actively and autonomously seek out and experiment with the game's mechanics. While such a "loose cannon" approach is often rather incompatible with the set curriculum that the pupils are required to learn (e.g., Skaug et al., 2020), it is certainly possible to argue for a more comprehensive approach to game-based learning. Were one to suggest subgoals for the pupils to reach while simultaneously encouraging free exploration and experimentation with the game, it is quite likely that the abovementioned example of using Eco in a classroom would yield a more promising outcome. As such, it is not necessarily the serious games themselves that are bad or ineffective – rather, it could very well be the *implementation and subsequent evaluation of the games* that are "missing the mark" in terms of how games generally function in terms of educating their players. It is therefore important, for anyone wishing to utilize games in sustainability education, to consider the following:

1. *Relevance of the game*

The first point of consideration should be the overall *relevance of the game to the educational context.* Although serious games tend to be explicitly tailored toward teaching skills that are applicable to real-life settings (Peters et al., 1998), there are often few boundaries as to how and when games can be used for educational purposes – even commercially available "nonserious" games. Much of a game's educational properties rely on the creativity and innovation on behalf of the educator or game administrator. Games such as *Minecraft* (Mojang., 2009) and *Papers, Please* (Pope, 2013), for example, can be used to teach about complex topics such as engineering and dystopian immigration policies, respectively – albeit with the creative liberties that might be expected from a game. As such, there are likely numerous games that can be utilized to teach about environmental topics that were not originally meant to be used for such purposes, again depending on the creativity of the educator. The Oil Springs scenario of Settlers of Catan (Assadourian & Hansen, 2011), for instance, is a good example of how one might include topic-specific components to an already existing game to use it for educational purposes and, subsequently, make it relevant to the field of environmental education.

2. *Aesthetics and mechanics of the game*

The original definition of the term "serious game," as we explored previously, is a game that seeks to do *more* than just entertain its players (Abt, 1970). However, this does not in any way imply that the field of serious games design should completely overlook the entertaining aspect of games – especially because a boring game is often considered unplayable (Sweetser & Wyeth, 2005) and, therefore, unsuitable for educational purposes. *Careful considerations of how the game looks, plays, and feels* are therefore needed to produce serious games that motivate, engage, and immerse their players into the game world. As serious games are often the subjects of low levels of graphical quality (Rice, 2006), they might be overlooked in favor of games with more attractive visual presentations. Furthermore, serious games that feature awkward controls or difficult and vague mechanics are likely to cause their players to spend more time learning how the game works rather than focusing on what it is designed to teach. By contrast, a good game should follow *Bushnell's Law*[2] of *being easy to learn but difficult to master* – i.e., it should carefully guide the player into the game world and allow them to establish a clear understanding of how it plays before clear challenges and frustrating obstacles can be introduced.

3. *Familiarization with the game*

Once a game has been chosen for environmental education purposes, it is crucial to *gain at least a basic understanding and familiarity of how the game operates.* If a game is chosen for its perceived ability to teach about very specific topics, it is important that the educator plays the game until they feel comfortable that the game

[2] https://en.wikipedia.org/wiki/Bushnell%27s_Law

truly teaches what it intends to teach. Furthermore, engaging with the game from an early stage will provide valuable information on how the game controls, looks, and feels. These are all central game enjoyment factors (Sweetser & Wyeth, 2005) and are likely to influence the degree to which the game is perceived as immersive and engaging in an educational setting (Fjællingsdal & Klöckner, 2017). Additionally, some games require in-depth familiarization for their true message to become clear. One such example is Brenda Romero's game *Train* which, at first glance, seems rather inconspicuous. The player is, simply enough, tasked with placing people – represented by yellow pegs – onto a train. However, at a later point in the game, the train's destination is revealed to be the infamous concentration camp Auschwitz, with the intention of causing an emotional shift from the joy of winning the game to the realization that one is an imaginary complicit in one of the darkest chapters of human history (Takahashi, 2013). This is also an excellent example for how disruptive events within a game context can create a (in this case painful) moment of reflection on following instructions and complicity.

4. *Complexity and length of the game*

Most games today have target audiences. Some games are tailored toward younger players, featuring less complexity and more basic game mechanisms, whereas other, more sophisticated games are made for older and more experienced players. *A game's degree of complexity will determine to what degree it is suitable for a specific audience and, by extension, to what degree the game educates its players.* Fate of the World (Roberts, 2011), for example, is a sophisticated environmental game developed for older, experienced players who can micromanage several different in-game elements at once. By contrast, the now defunct browser-based Energy Saving Game (https://www.olliesworld.com/club/gamehouse.htm) was tailored toward younger, less experienced players, required far less micromanagement, and had a considerably shorter playtime (Klöckner, 2015, p. 198). Furthermore, it is also important to consider the *complexity of the game itself* in addition to the complexity of the *game's topic*. There are limits as to how many cognitive resources one can dedicate to environmental topics (Gifford, 2011), and the same applies to how many resources one can be expected to dedicate to learning game mechanics. In addition to the game's level of complexity, it is also useful to consider how long it takes to play through it. Although there are several examples of environmental games that only take anything from a few minutes (e.g., the abovementioned Energy Saving Game) to a couple of hours to play (e.g., Keep Cool or Fate of the World), others are significantly more time-consuming and sophisticated. Eco, for instance, can take as much as 30 real-life days to "complete" (Fjællingsdal & Klöckner, 2019), and even then, the game continues as long as the player chooses to keep playing. In the case of board games, CO_2 (Lacerda, 2012) also requires significant time investment both in terms of setting up the board, learning how the game functions, and playing it – largely due to the level of complexity and the high level of difficulty for new or inexperienced players (Klöckner, 2015, p. 203).

5. Debriefing after the game

A common "misconception" about serious games is that the game itself is the only joint in the game-based learning process. Although the design and presentation of the serious game is certainly important, research overall suggests that it is just as important for educators to conduct postgame *debriefing* – or "the occasion and activity for the reflection on and the sharing of the game experience to turn it into learning" (Crookall, 2010). By collectively reflecting on the gameplay experience, it is possible to share experiences, clarify misconceptions surrounding the game's presentation of core topics, brainstorm pro-environmental ideas based on the game's framework, invent clever real-life solutions for problems arising in the game world, and so on. As such, serious games (environmental ones included) therefore become not a replacement for conventional teaching methods but rather a complementary supplement (Meya & Eisenack, 2018). However, as there is yet little empirical research on the learning effectiveness of serious environmental games (Hallinger et al., 2020), many of the positive effects of postgame debriefing still largely exist on a theoretical level. It is therefore crucial in future research on environmental games that debriefing sessions are conducted, evaluated, and subsequently improved based on this emerging empirical foundation, a feature that the games discussed here share with many other disruptive interventions presented in this book.

6. Acknowledge the barriers of environmental game implementation

Lastly, and perhaps most importantly, is the understanding of existing barriers and issues with using games in specific contexts. While this chapter has provided a basic introduction to some of the more generalized barriers to game-based learning, there exists a wealth of empirical research emphasizing game-specific and context-specific barriers as well. In some cases, for example, the intended audience for the game one wishes to use might not have the required degree of literacy with games to play it effectively (Illingworth, 2020). In other cases, the game can be difficult to integrate properly with the existing student curriculum, or they might simply take too much time to play (Fjællingsdal & Klöckner, 2019). Games are, therefore, by no means an end-all solution to every environmental education issue. Simultaneously, however, they can work well in combination with other learning methods (Meya & Eisenack, 2018). It is therefore important to consider the context in which the serious environmental game is to be implemented and whether the positive sides of using one outweigh the limitations and barriers.

Following these suggested guidelines will give educators the necessary tools with which to both (1) understand the inner and outer workings of the environmental game they wish to use and (2) make them able to evaluate to what degree the game has taught its players about the topic in question using postgame debriefing. The guidelines are suitable for both video- and board-based games and are built on core concepts from previous empirical research on the use of serious games in sustainability education (Fjællingsdal & Klöckner, 2017, 2019, 2020; Sandbrook et al., 2015). As such, they will hopefully be utilized by future educators and researchers within the domains of sustainability, ecology, and the environmental sciences who wish to explore the possibilities of – as the title of this chapter suggests – using entertaining approaches to instigate changes in pro-environmental knowledge.

References

Abt, C. C. (1970). *Serious games*. University Press of America.

Ajzen, I. (1991). The theory of planned behavior. *Organizational Behavior and Human Decision Processes, 50*(2), 179–211.

Assadourian, E., & Hansen, T. (2011). *Catan scenarios: Oil springs [board game]*. Catan GmbH.

Bartsch, A., & Viehoff, R. (2010). The use of media entertainment and emotional gratification. *Procedia Social and Behavioural Sciences, 5*, 2247–2255.

Bogost, I. (2010). *Persuasive games. The expressive power of videogames*. MIT Press.

Boomsma, C., Hafner, R., Pahl, S., Jones, R. V., & Fuertes, A. (2018). Should we play games where energy is concerned? Perceptions of serious gaming as a technology to motivate energy behaviour change among social housing residents. *Sustainability, 10*(6), 1–18.

Brown, E., & Cairns, P. (2004, April). *A grounded investigation of game immersion*. Paper presented at CHI 2004, Vienna, Austria.

Burn, A. (2008). The case of rebellion: Researching multimodal texts. In J. Coiro, M. Knobel, C. Lankshear, & D. J. Leu (Eds.), *Handbook of research on new literacies*. Lawrence Erlbaum Associates.

Bushman, B. J., & Anderson, C. A. (2001). Violent video games and hostile expectations: A test of the general aggression model. *PSPB, 28*(12), 1679–1686.

Crookall, D. (2010). Serious games, debriefing, and simulation/gaming as a discipline. *Simulation & Gaming, 41*(6), 898–920.

Dimopoulos, K., & Koulaidis, V. (2003). Science and technology education for citizenship: The potential role of the press. *Science Education, 87*(2), 241–256.

Egenfeldt-Nielsen, S., Smith, J. H., & Tosca, S. P. (2013). *Understanding video games*. Routledge.

Eisenack, K., & Petschel-Held, G. (2004). *Keep cool [board game]*. Spieltrieb.

Ermi, L., & Mäyrä, F. (2005, June). *Fundamental components of the gameplay experience: Analysing immersion*. Paper presented at DiGRA 2005 conference: Changing views – Worlds in play, Vancouver, Canada.

Fazio, R. H., & Zanna, M. P. (1978). On the predictive validity of attitudes: The roles of direct experience and confidence. *Journal of Personality, 46*(2), 228–243.

Fjællingsdal, K. S., & Klöckner, C. A. (2017). ENED-GEM: A conceptual framework model for psychological enjoyment factors and learning mechanisms in educational games about the environment. *Frontiers in Psychology, 8*, 1–17.

Fjællingsdal, K. S., & Klöckner, C. A. (2019). Gaming green: The educational potential of eco – A digital simulated ecosystem. *Frontiers in Psychology, 10*, 1–13.

Fjællingsdal, K. S., & Klöckner, C. A. (2020). Green across the board: Board games as tools for dialogue and simplified environmental communication. *Simulation & Gaming, 51*(5), 632–652.

Galarneau, L. (2005). Authentic learning experiences through play: Games, simulations and the construction of knowledge. In *Proceedings of DiGRA 2005 Conference: Changing views – Worlds in play*.

García-Barros, L., García-Barros, R., Cruz-Morales, J., & Smith, J. A. (2015). When death approaches: Reverting or exploiting emergent inequity in a complex land-use table-board game. *Ecology and Society, 20*(2), 13.

Gifford, R. (2011). The dragons of inaction. Psychological barriers that limit climate change mitigation and adaptation. *American Psychologist, 66*(4), 290–302.

Greeley, A. (1993). Religion and attitudes toward the environment. *Journal for the Scientific Study of Religion, 32*(1), 19–28.

Greitemeyer, T. (2013). Beware of climate change skeptic films. *Journal of Environmental Psychology, 35*, 105–109.

Hallinger, P., Wang, R., Chatpinyakoop, C., Nguyen, V.-T., & Nguyen, U.-P. (2020). A bibliometric review of research on simulations and serious games used in educating for sustainability, 1997–2019. *Journal of Cleaner Production, 256*, 1–15.

Hansen, A., & Machin, D. (2013). Researching visual environmental communication. *Environmental Communication: A Journal of Nature and Culture, 7*(2), 151–168.
Hines, J. M., Hungerford, H. R., & Tomera, A. N. (1987). Analysis and synthesis of research on responsible environmental behaviour: A meta-analysis. *The Journal of Environmental Education, 18*(2), 1–8.
Huckle, J. (1993). Environmental education and the national curriculum in England and Wales. *International Research in Geographical and Environmental Education, 2*(1), 101–104.
Illingworth, S. (2020). Creative communication – Using poetry and games to generate dialogue between scientists and nonscientists. *FEBS Letters, 594*, 2333–2338.
Jennett, C., Cox, A. L., Cairns, P., Dhoparee, S., Epps, A., Tijs, T., & Walton, A. (2008). Measuring and defining the experience of immersion in games. *International Journal of Human Computer Studies, 66*(9), 641–661.
Klöckner, C. A. (2015). *The psychology of pro-environmental communication. Beyond standard information strategies*. Palgrave Macmillan.
Koehler, M.J., Greenhalgh, S.P., & Boltz, L.O. (2016, March). *Here we are, now entertain us! A comparison of educational and non-educational board games*. Paper presented at SITE 2016, Savannah, United States.
Kollmuss, A., & Agyeman, J. (2002). Mind the gap: Why do people act environmentally and what are the barriers to pro-environmental behaviour? *Environmental Education Research, 8*(3), 239–260.
Lacerda, V. (2012). *CO2 [board game]*. Giochix.it.
McCright, A. M., Xiao, C., & Dunlap, R. E. (2014). Political polarization on support for government spending on environmental protection in the USA, 1974–2012. *Social Science Research, 48*, 251–260.
McGonigal, J. (2012). *Reality is broken. Why games make us better and how they can change the world*. Vintage Books.
Meya, J. N., & Eisenack, K. (2018). Effectiveness of gaming for communicating and teaching climate change. *Climatic Change, 149*, 319–333.
Mojang. (2009). *Minecraft (PC version) [video game]*. Mojang Studios.
Moser, S. C. (2010). Communicating climate change: History, challenges, process and future directions. *WIREs Climate Change, 1*(1), 31–53.
Myers, T. A., Maibach, E. W., Roser-Renouf, C., Akerlof, K., & Leiserowitz, A. A. (2013). The relationship between personal experience and belief in the reality of global warming. *Nature Climate Change, 3*(4), 343–347.
Nacke, L., & Lindley, C.A. (2008, January). *Flow and immersion in first-person shooters: measuring the player's gameplay experience*. Paper presented at the 2008 conference on future play: Research, play, share, future play 2008, Toronto, Ontario, Canada.
Nakamura, J., & Csikszentmihalyi, M. (2014). The concept of flow. In M. Csikszentmihalyi (Ed.), *Flow and the foundations of positive psychology*. Springer.
Ouariachi, T., Olvera-Lobo, M. D., & Gutiérrez-Perez, J. (2018). Serious games and sustainability. In W. L. Filho (Ed.), *Encyclopedia of sustainability and higher education*. Springer.
Peters, V., Vissers, G., & Heijne, G. (1998). The validity of games. *Simulation & Gaming, 29*(1), 20–30.
Pope, L. (2013). *Papers, please (PC version) [Video game]*. 3909 LLC.
Reckien, D., & Eisenack, K. (2013). Climate change gaming on board and screen: A review. *Simulation & Gaming, 44*(2–3), 253–271.
Reis, R. (1999). Environmental news. Coverage of the earth summit by Brazilian newspapers. *Science Communication, 21*(2), 137–155.
Rice, J.W. (2006). *New Media Resistance: Barriers to Implementation of Computer Video Games in the Classroom*. Paper presented at American Educational Research Association, 2006 annual meeting San Francisco, CA.
Roberts, I. (2011). *Fate of the world [PC game]*. Red Redemption.
Salen, K., & Zimmerman, E. (2004). *Rules of play. Game design fundamentals*. The MIT Press.

Sandbrook, C., Adams, W. M., & Monteferri, B. (2015). Digital games and biodiversity conservation. *Conservation Letters, 8*(2), 118–124.

Schell, J. (2008). *The art of game design: A book of lenses*. CRC Press.

Sheppard, S. R. J. (2012). *Visualizing climate change. A guide to visual communication of climate change and developing local solutions*. Routledge.

Skaug, J. H., Husøy, A., Staaby, T., & Nøsen, O. (2020). *Spillpedagogikk. Dataspill i undervisningen [game pedagogy. Computer games in lecturing]*. Fagbokforlaget.

Skukan, R., Borrell, Y. J., Ordás, J. M. R., & Miralles, L. (2020). Find invasive seaweed: An outdoor game to engage children in science activities that detect marine biological invasion. *The Journal of Environmental Education, 51*(5), 335–346.

Smith, B. J., Ferguson, C., McKenzie, J., Bauman, A., & Vita, P. (2002). Impacts from repeated mass media campaigns to promote sun protection in Australia. *Health Promotion International, 17*(1), 51–60.

Stoknes, P. E. (2017). *Det vi tenker på når vi prøver å ikke tenke på global oppvarming [what we think about when we try not to think about global warming]*. Tiden Norsk Forlag.

Strange Loop Games. (2020). *Eco [PC game]*. Strange Loop Games.

Sweetser, P., & Wyeth, P. (2005). GameFlow: A model for evaluating player enjoyment in games. *ACM Computers in Entertainment, 3*(3), 1–24.

Takahashi, D. (2013, May 11). *Brenda Romero's Train board game will make you ponder*. Retrieved 24 November 2020 from https://venturebeat.com/2013/05/11/brenda-romero-train-board-game-holocaust/

van Rooij, A. J., Schoenmakers, T. M., Vermulst, A. A., van den Eijnden, R. J. J. M., & van de Mheen, D. (2010). Online video game addiction: Identification of addicted adolescent gamers. *Addiction, 106*, 205–212.

Weibel, D., & Wissmath, B. (2011). Immersion in computer games: The role of spatial presence and flow. *International Journal of Computer Games Technology, 2011*, 1–14.

Wu, J. S., & Lee, J. J. (2015). Climate change games as tools for education and engagement. *Nature Climate Change, 5*(1), 413–418.

Yee, N. (2014). *The Proteus Paradox. How online games and virtual worlds change us – And how they don't*. Yale University Press.

Chapter 6
Provotyping and Eco-visualization: Disruptive Workshops to Question Status Quo and Co-design Solutions

Erica Löfström and Kristoffer S. Fjællingsdal

6.1 Introduction

Our societies are based on consumption of products and services that are often designed by engineers or product designers within the tight framework of what is considered possible and without consulting the citizens who are supposed to use them afterward (e.g., Morris, 2016). In addition, consumers are expected to make aware choices between the available products and services in their everyday lives. This chapter presents a *strategy for designing* breaking with this tight framework, opening for exploring extreme ideas and thus harvesting the creative power of citizens in co-design processes. It also exemplifies how *provotypes* can be used to spark discussions at a societal level, hence challenging the status quo – including the choices available for consumers – and what this may entail with regard to our common society and future.

6.2 Provotyping: Different Levels and Uses

Provotypes are short for provocative prototypes (Boer & Donovan, 2012). As such, they are a variation of prototypes, but their use is clearly different. While prototypes are intended to become actual products or services, provotypes have another purpose; they are disruptive and designed especially to provide alternative and thought-provoking experiences (Löfström et al., 2020) to those that utilize them. Similar to prototypes, provotypes may be rather simplistic, such as in the case of using fake screenshots to test an app while someone discretely switches to new screenshots depending on the choices the test person makes (often referred to as "Wizard of Oz" – usability test situation (Hosseini-Khayat et al., 2009)). They may, however, also be highly advanced, such as in the case of fully functioning artificial Intelligences

C. A. Klöckner, E. Löfström, *Disruptive Environmental Communication*, Psychology and Our Planet, https://doi.org/10.1007/978-3-031-17165-9_6

(AIs) or home robots. The use of prototypes is a very common method in design and co-design of products and services (Yang, 2005), which may be used in all stages of the design process, while provotypes have mainly been used in research contexts.[1] Provotypes are intentionally designed to provoke resistance and reflection, and while the results may well result in ideas for products/services and form the basis of prototypes, this is not their main purpose (Boer & Donovan, 2012). However, the use of provotypes requires thorough preparation not only with regard to the provotype itself but also in the way it is presented and in which context. The need is to successfully rock barriers allowing for the exploration of possibilities without triggering too many of our innate psychological defense mechanisms (Cramer, 2006). We will present some examples on the use of provotypes in different settings to illustrate both possibilities and pitfalls in using them for the purpose of "questioning status quo and co-designing solutions together" with different stakeholders.

6.3 Radical Change Is Challenging

As described and elaborated on in Chap. 2, radical change is not something that comes easily for humans. There are a number of psychological mechanisms and social psychological processes in place to prevent us from changing our current way of life, such as the *status quo bias* (Samuelson & Zeckhauser, 1988) that makes us prefer the least change possible and to stick to what we know (Arkes & Blumer, 1985; Kahneman & Tversky, 1982), *cognitive dissonance* (Festinger, 1962), *subconscious barriers* to behavioral change (Gifford, 2011), or *self-perception* (Bem, 1972) that makes us strive to be consistent in our choices and behavior, *habits* and behavioral routines that are usually not even reflected upon (Aarts et al., 1998; Verplanken et al., 1994; Klöckner & Matthies, 2004; Russell et al., 2017), *social norms* and *values* (Cialdini & Jacobson, 2021) that dictate what is considered "normal," and social and individual *identities* which represent the need of individuals to define who they are (Stets & Burke, 2000). A successful provotyping co-design workshop will challenge one or more of these mechanisms while being thematically playful and interesting in a nonthreatening way. The purpose of disruptive communication itself is to challenge the actions, but not the individuals.

Therefore, before willfully challenging and provoking the participants, the experimenter needs to open up a space for reflecting upon and discussing potential solutions to an issue in a safe environment. Opening up a safe place for dreaming and imagining is key for successful provotyping workshops. One way of doing this is that the experimenter shows that he or she is willing to share their wild and unrealistic ideas with the group and to demonstrate that this has no negative consequences in this setting. For this, we have found that taking inspiration from the "yes, and" rule of improvisational theater is a useful tool (Nachmanovitch, 1990). The

[1] https://medium.com/@thestratosgroup/moving-from-prototyping-to-provotyping-cedf42a48e90

rule is that whatever someone else says, your response should be based on a "yes, and" approach as opposed to a "no, but" approach. The logic is that, when two or more discussants follow each other's ideas and comments – playing along with each other's commentary instead of dismissing it – they build on and multiply each other's ideas. By applying the "yes, and" rule, you encourage and build on a shared creative flow. As an experimenter, you need to convey the message that all ideas are to be treated as "seeds" that may eventually grow into more concrete ideas. Specifically, in the workshop introduction setting, the experimenter may share an idea that they have come up with that initially seemed erratic or unrealistic but was further developed into a successful concept, project, service, or product thanks to the "yes, and" approach of their colleagues. As a researcher it is perhaps easier to use a project idea as an example, as they usually take a significant amount of time to develop, fund, and realize. It may also be good to share examples of cognitive dissonance in everyday life to avoid that the participants feel the need to defend their behavior or choices in their life during the workshop – such as by illustrating how we, as humans, tend to justify the choices we make (e.g., choosing to use a car to drive to one's workplace instead of using public transportation, due to how cars can be more practical, accessible, and comfortable than a bus).

6.4 Provotypes as Tools for Radical Change

As mentioned earlier, provotypes can be used on different levels and for different purposes. For instance, they can be used to find out where the limits to what is perceived as morally or socially acceptable behavior or technology go, challenging both social and individual norms, values, and identities. For this particular type of workshop, you could, for example, use an actor to play the role of an artificial human or robot – i.e., our provotype. The experimenter could then tell the participants of the workshop that they are going to interact with the latest prototype of a social support artificial intelligence robot that will be launched later that year, if the human-robot interaction proves to be successful in the current experiment. The participants will test interacting with this robot and help determine whether it can be safely and without serious complications used in care environments. The actor serves as the workshop "provotype."

The abovementioned provotype may help us determine what kind of interaction with an AI is considered socially desirable and what it should be able to do in real-life, everyday care environments. By exemplifying which tasks can be carried out by our provotype AI, we can by extension find out which challenges and barriers the experiment participants face in terms of adopting it as an acceptable tool or solution. For instance, how will the participants react to the AI feeding them, washing their hair, or shaving them? Furthermore, what might happen if the provotype begins to show emotional involvement? What if it smiles, laughs, or cries? For inspiration, the science fiction literature is packed with examples of AIs getting a crucial role in relation to humans, spanning from the 1960s and up until today (see, for instance, Ezpelata & Segarra, 2017: 64–68; Hanley, 1997).

The possibilities are, of course, endless. Besides being a very exciting experiment, this exercise may reveal not only *if* it is problematic to have AIs carrying out tasks and helping us in our everyday life but *in which way* this is problematic. Perhaps especially interesting is the issue of the participants not being able to distinguish between a human being and a machine. The results of such an experiment may be interesting for researchers from many different disciplines – including philosophy and psychology – and may give valuable insights into, for instance, how people understand the world and the basis of human interaction. In essence, provotyping is there used to determine the limits to what is socially and technologically acceptable when it comes to robots replacing humans in care environments. However, for our co-design workshops, this setting may also be useful in determining more *specifically* how a social service robot *should* be constructed and what we would like for it to be able to do and not to do, hence opening up for reflections on the care services for future generations in light of the problem with aging populations (Bengtsson & Scott, 2011) and the expected extreme shortage of care personnel in wealthier countries.[2] This may well result in the co-development of both products and services for care environments.

Continuing with the experiment with our care provotype AI, we could now add to the experiment the information that the person receiving the care has dementia and does not realize that their caregiver is an AI. What if the patient starts to trust the AI and talks to it in confidence? Is this then positive or negative and – if so – why?

The workshop can then focus on how this should be handled and whether it should be avoided. If so, what are the traits that need to be there in order to "reveal" or remind the caretaker that this is a robot?

Using the example of an AI caregiver as a provotype in a co-creative workshop setting, we have seen that it is possible to explore different scenarios and to add or change conditions during the workshop. However, whatever the focus of the workshop is, it may be necessary to remind the participants of the "safe space" and on upholding the "yes, and" approach during the workshop.

To conclude, the use of an AI caregiver provotype in a workshop setting is, of course, not to introduce the robot as a prototype but to spark discussion using it as a *provotype*. This example was chosen partly because it is a versatile example but also because AIs are expected to become part of future case environments and thus at least partly solving a major problem facing humankind.

If we instead present the same AI/robot as a *personal moral compass-resource* that will help people act more environmentally friendly, the discussions may really take off. How would the participants react if this *moral eco-bot* provides us with information on the carbon footprint of products and services? That would probably be okay. But what if it suggests other available alternatives instead? What if it insists

[2] https://www.euro.who.int/__data/assets/pdf_file/0004/64966/E92560.pdf

on another more sustainable product or service? What if it gives an emotional reaction to bad choices, crying, or screaming at you? What if it helps you uphold, say a regulated personal carbon quota? And, to push it even further, what if it reports you to the authorities should you exceed your quota? Perhaps luckily, such AIs do not yet exist and are perhaps not something we would even wish for or consider, but this eco-bot example effectively illustrates the potential of using provotypes to spark discussions on environmental sustainability issues and behaviors, which may be used as a starting point for co-development of new products, services, or practices. We refer to this type of provotypes as eco-provotypes.

6.5 Provotypes for Co-creating Our Sustainable Future?

Provotyping can help us to speculate about potential clues to a more sustainable future (Dunne & Raby, 2013). Used this way, provotypes can expose and embody tensions that surround a field of interest to support collaborative analysis and collaborative design explorations. The concept of provotyping is related to critical design (Dunne & Raby, 2012), which is a strategy to cultivate so called critical sensitivity. This involves using design to stimulate critical thinking and group-based discussions for questioning the norms of society. Using tools to support creative working groups (Johnson & Johnson, 2009, p. 20), it is possible to generate two-way communication practices in which the members can utilize their expertise as well as feedback from others in order to refine basic concepts and ideas that would otherwise be considered absurd or impossible to implement.

However, if understood as eco-provotypes, they are not general or open but revolve around pro-environmental values to help focus the discussions thematically and to "frame" them into larger or smaller contexts of sustainable resource use. In the following examples we will demonstrate, how these values are supported by the use of *eco-visualizing* techniques. The following definition has been used for eco-visualization:

> Communicating (eco-) system interdependencies with the intent of achieving awareness, flexibility and (societal) debate (Löfström & Svanæs, 2017).

However, we wish to point out that even if eco-visualizations are helpful in the process and we often use them in our practice, these pro-environmental values do not need to be supported by actual eco-visualization techniques and actual data in order to make a provotype an eco-provotype. It is enough that the provotype includes an outspoken message or direction toward *pro-environmentalism.*

In the following section, we will use examples from previous workshops where we have co-designed and developed concepts, services, and/or products together with end users in the fields of environmental sustainability and sustainable resources using eco-provotypes.

6.6 Provotyping the Smart Grid: Questioning the Electricity System

The smart grid application provotype was used in workshops during the planning stage of developing communication with end users that were introduced with a smart metering system as part of a so-called smart grid.[3] This smart metering system would communicate with the end users and encourage them to save electricity – and to move the use to times of the day when electricity is not scarce (usually referred to as load-shifting). In order to determine what would be interesting and relevant information for the users to receive, mock-up messages (early-stage or nonfunctioning prototypes/provotypes) as part of a *hypothetical* (non-existing) app solution were used to kick-start discussions. The messages were informing about problems with the delivery of electricity (no access or technical distribution issues) and asked the participants to willingly use less electricity in order for others to have access.

The study (Löfström, 2014a) revealed which type of data the users were interested in getting from the provider, i.e., price of electricity, technical issues with the smart grid, and environmental issues. So far, these mock-ups were more of an early-stage prototype type. However, we also tried out specifying different groups as being the ones who would benefit from their decision to "give" their electricity to others who are part of the same energy distribution net (Palensky & Kupzog, 2013).[4] This is where the mock-ups became disruptive and the provotyping began. For instance, we tested a mock-up message where the users were asked whether they would be willing to postpone or omit using electricity altogether when there was a shortage of power in order to secure access for senior citizens at a care home. These mock-ups did not only add to the discussions on which smart metering data they would like to receive or what might motivate them to change their consumption patterns. As it turned out, it also kick-started discussions of a broader form, getting into both political and global issues. With the use of provotypes, this initially local experiment became a discussion ground for regional-, national-, and – all the way up to – global-level topics (Löfström, 2014a). After a while the discussions became quite heated, and participants started to become critical toward the status quo of the electricity production and distribution system in Norway.

What is likely to have provoked these discussions was that the mock-up app messages presented their separate household as part of a larger system of production and distribution of electricity. In essence, the participants likely gained a shared understanding of how the electricity/energy production and distribution net functions, based on feedback from the other parties engaged in the debate. Receiving reactions and feedback from a group in a debate setting is known to provide a rich arena for self-understanding and reframing one's experiences of a subject (Johnson & Johnson, 2009, p. 511), which, in this case, revolved around the perception of

[3] A smart grid can be defined as the infrastructure system from the production of electricity, onward to the balanced distribution of it to different users.

[4] https://en.wikipedia.org/wiki/Smart_grid

electricity. As a consequence, they also likely gained a deeper understanding of their own role in the production and consumption of household electricity. Hence, their own choice to refrain from using electricity for altruistic reasons was put into a larger societal context, and these ethical/moral dilemmas became a starting point for discussions on societal issues rather than individual households.

As mentioned earlier, to visualize the consequences of resource use with the intent to promote sustainable behavior, decision-making, and/or attitudes is called "eco-visualization" or "eco-feedback" (Löfström & Svanæs, 2017), and this was included in the app. However, in this mock-up app provotype, the visualization properties not only focus on individual behavior but also individual behavior in a systemic context. This can be referred to as systemic eco-visualizations. In our abovementioned workshop setting, this provotype with systemic eco-visualization elements provided an arena for discussions on a systemic level. It seems that using provotypes with successful systemic eco-visualizations allows for more in-depth, unrestrained discussions surrounding controversial or complex topics. We have successfully travelled from the private sphere and into the public, which means we can discuss not only changes on individual or household level but also the systemic conditions surrounding individual consumption patterns. This becomes especially important in relation to the aforementioned issue of finite resources (Schuurman et al., 2022), as individuals systematically tend to consider themselves practitioners of "the business of living" rather than active consumers (Lorek & Vergragt, 2015). Additionally, provotypes may give rise to discussions that entail policy-relevant and political aspects. As the lead researcher in two different groups of electricity consumers, it was somewhat overwhelming to see the strong engagement that rose from the debate. To illustrate this, the following quotes from the study are presented (Löfström, 2014a):

> I think it would be a good idea to have a more open approach in relation to information.

> I would suggest we create a society where we actually are playing together, on the same team.

> We are able to draw our own conclusions on this matter. It seems they [as in decision-makers] do not trust us to be smart enough.

The results of these workshops, in which we used eco-provotypes, turned out to be more potent than initially theorized, and the subsequent discussions should be explored further. To conclude, it seems that using "systemic eco-provotypes" does not only allow for the co-creation of ideas and solutions in a workshop setting but also contributes to successfully lifting the discussions to a societal systemic level, which certainly deserves a larger audience and impact than simply as part of a research experiment. Following up on this project, we would encourage structuring projects that use "systemic eco-provotypes" as a way of involving citizens in policymaking and even political discussions – effectively creating a citizen science approach to the sustainability debate (see Kullenberg & Kasperowski, 2016).

6.7 An Eco-visualization Tower in a Sustainable Neighborhood

In another project that was related to the planning and development of a carbon-neutral residential neighborhood, creative workshops were carried out with potential future residents (Löfström, 2014b). In this project, the possibility of using public visualizations of the consumption, represented by carbon foot-printing and CO_2 emissions for each household of a neighborhood systemic context, was used as part of a provotyping concept. The participants were shown an illustration of an "eco-visualization tower" (see Fig. 6.1) which, among other things, makes the private households consumption patterns observable in a public setting. The system here was limited to a specific neighborhood. The most intriguing part of this study revolved around discussions on whether consumption patterns should be made public, as the consequences of overconsumption actually do concern us all at a systemic (global) level. The co-designed actual solutions for the neighborhood were pretty much what may be expected, but the discussions culminated in a common need among the participants to signal their identity as a sustainable neighborhood and to clearly deviate from that of a "normal" neighborhood. In addition to the sharing of household tools, intergenerational social activities, and cooperating with others in cooking, growing food, doing laundry, and walking children to school, the identity markers turned out to be essential. The group came up with ideas for a local mini-rail for mobility inside the area as they soon decided that keeping the neighborhood car free would be fundamental. They also decided that they wanted a version of the eco-visualization tower to signal the low-carbon identity and goals both internally and to the outside world.

This project was an early attempt to use a provocative concept to spark discussions, debates, and enabling creative co-designing processes. In addition, this study was performed by an early-stage researcher and the methodology of using a

Fig. 6.1 The eco-visualization tower. Illustration: team SLA, Brøset parallel assignment

combination of systemic eco-visualizations together with provotypes was still emerging (Löfström, 2008). While the use of provotyping in combination with (systemic) eco-visualizations was present in the research, this was not yet fully articulated. In hindsight, the discussions could probably have been lifted to include more general societal issues, such as conditions and restrictions for neighborhood production of energy or circular economy, but the project nevertheless did provide some important input regarding how to support carbon-neutral lifestyles at a neighborhood level.

6.8 Public Visualization of Private Consumption Choices

The idea of a tool to help the shoppers make decisions between different products was developed as part of another study (Löfström & Peterssen, 2011). Once a person's impact on the environment is determined, for example, by comparing the contents of their shopping trolley with that of the "average" consumer, this information can then be visualized for the shopper. This would not only provide price comparisons between different goods but also their relative environmental impacts. Although the necessary calculations for developing such an app solution did not exist at the time of the study, the possibility of using such a solution was explored and became part of a co-designing process.

Another version of this conceptual tool, the *morally concerned teddy bear*, was introduced as well. It would cry if a product on the shop shelf was non-sustainable, be happy if it was, and featured a cashier solution called *reflective surface* that would reveal the product's carbon footprint.[5] Focus group discussions contributed to a fruitful co-creation process on developing these tools and concepts further, and today apps like the one presented as a mock-up to the participants do exist and are being used worldwide (Weber, 2021).

However, from a research perspective, the perhaps most interesting discussions emerged when a confrontational element was introduced to the groups. The idea was that the mock-up tool would communicate with the shopping trolley, and the trolley would signal how sustainable the content of the trolley was by using green and red lights that would be visible for other people in the store. The "morally concerned teddy bear" would now cry loudly and object to the consumer's choices. In addition, the conceptual *Sustainabag* was introduced into the discussion – a shopping bag that would signal on its surface to what extent the consumer had made non-sustainable choices. If the carbon footprint of the bag's contents was high, the handles would automatically become longer and impractical, which was thought to add to the consumer's feeling of discomfort. When adding the factor of public exposure to the prototypes and concepts, they instead became *provotypes*. The concept

[5] The reflective surface and the teddy bear were *prototypes* developed by the Interactive Institute, Stockholm, Sweden.

of making this private information public was considered both more potentially effective, but also – for some participants – worrying. It was clear that the level of public access of the information was important.

The concept of public exposure in the store and possibly to people meeting you on your way home was met with two different fractions among the focus group participants. Surprisingly, it did not have to do with any aspects of privacy or personal integrity but rather emanated from whether the focus group participants perceived themselves as more or less likely to act in an environmentally sustainable way. The ones who pictured themselves as doing the "right thing" said that they would be proud if others could see their behavior, while the ones picturing themselves doing what was considered "less positive" said they would be ashamed or at least embarrassed. Interestingly, the concept of a public display of personal choice was not dismissed as unrealistic or unethical but resulted in discussions on what should be public and what should be private (Löfström & Peterssen, 2011). The consensus in one of the groups was that personal choices have far-reaching consequences. This represents a potential shift in perspective on private consumption and choices as no longer belonging solely to the personal sphere.

6.9 Discussion and Further Research

The use of early-stage provotypes (mock-ups) has, from a methodological aspect, a lot in common with "critical design" and "design noir" (Dunne & Raby, 2012). In critical design, early-stage provotypes are used as a strategy to cultivate what Dunne and Raby call *critical sensitivity*. This involves using design to stimulate critical thinking and people questioning the norms of society. However, while critical design simply aims at questioning of norms in general, eco-*provotyping* also involves an ambition of change and of finding new and more sustainable solutions together. This means that the eco-provotypes used in these studies are not neutral. They are designed to have an inscribed and outspoken message or direction. In this sense, the "combining provotypes with eco-visualizations" method shares considerable overlap with artistic approaches to communication.

As our examples have shown, all experiments spurred discussion and reflections, although on various topics and on different levels with regard to resource use. The most potent experiment was perhaps the smart grid mock-up app (Löfström, 2014a) which boosted discussions on how electricity is produced and distributed and on what information is available or unavailable for the consumers. In this context, it was clear that the participants felt that they had been excluded from relevant information and that they felt underestimated. They wanted to be part of important decisions with regard to electricity in a broader sense, at a societal level. As this experiment clearly showed, this kind of disruptive communication – early-stage eco-provotype mock-ups of a smart grid – can be used to spark discussions on a societal systemic level as well as providing an access point to discuss a more sustainable future. However, the potency of the method calls for a clearer plan with regard to taking their engagement further into a context where they can take part in actual decisions. After all, what is the use of engagement and participation in co-creating a sustainable future if it does not influence the decisions we make as a society?

The visualization tower provoked a debate about the (public) nature of private consumption (Löfström, 2014b), while the grocery store provotypes (Löfström & Peterssen, 2011) provoked reflection for individual users concerning environmental consequences of private choices. However, the most surprising finding from these studies was related to the young adults' response to making individual choices public as part of the grocery shopping experience. In light of these young adults' attitudes, we now might wish to go back to our dubious AI-friend – the moral eco-bot – that may not seem at all as far-fetched as when it was introduced in this chapter.

All the presented studies are exploratory and were carried out with limited research funding. The results show that following up with further studies on how to use eco-provotypes represents a promising future research arena – especially within the realm of global morality and biospheric values, the eco-conscience of young individuals, and the possibility to lift the engagement to a level where it can influence the decisions we make as a society. Altogether, this may not only help finding solutions for our hopefully sustainable future but may also improve democratic empowerment by increasing public participation in debates at a societal level.

References

Aarts, H., Verplanken, B., & van Knippenberg, A. (1998). Predicting behaviour from actions in the past: Repeated decision making or a matter of habit? *Journal of Applied Social Psychology, 28*(15), 1355–1374.

Arkes, H. R., & Blumer, C. (1985). The psychology of sunk cost. *Organizational Behaviour and Human Decision Processes, 35*(1), 124–140.

Bem, D. J. (1972). Self-perception theory. *Advances in Experimental Social Psychology, 6*, 1–62.

Bengtsson, T., & Scott, K. (2011). Population aging and the future of the welfare state: The example of Sweden. *Population and Development Review, 37*(Suppl 1), 158–170. https://doi.org/10.2307/41762403

Boer, L., & Donovan, J. (2012, June). Provotypes for participatory innovation. In *Proceedings of the designing interactive systems conference* (pp. 388–397). ACM.

Cialdini, R. B., & Jacobson, R. P. (2021). Influences of social norms on climate change-related behaviours. *Current Opinion in Behavioural Sciences, 42*, 1–8.

Cramer, P. (2006). *Protecting the self: Defense mechanisms in action*. Guilford Press.

Dunne, A., & Raby, F. (2012). *Critical design FAQ*. Available: https://www.readingdesign.org/critical-design-faq. Accessed 5 Mar 2022.

Dunne, A., & Raby, F. (2013). *Speculative everything: Design, fiction, and social dreaming*. MIT Press.

Ezpelata, D., & Segarra, I. (2017). Transhumanism on artificial intelligence portrayed in selected science fiction movies and TV series. *MEDIC, 25*(1), 64–68.

Festinger, L. (1962). Cognitive dissonance. *Scientific American, 207*(4), 93–107.
Gifford, R. (2011). The dragons of inaction: Psychological barriers that limit climate change mitigation and adaptation. *American Psychologist, 66*(4), 290–302. https://doi.org/10.1037/a0023566
Hanley, R. (1997). *Is data human?: The metaphysics of star trek*. Basic Books.
Hosseini-Khayat, A., Ghanam, Y., Park, S., & Maurer, F. (2009). ActiveStory enhanced: Low-fidelity prototyping and wizard of oz usability testing tool. In P. Abrahamsson, M. Marchesi, & F. Maurer (Eds.), *Agile processes in software engineering and extreme programming. XP 2009* (Lecture notes in business information processing) (Vol. 31). Springer.
Johnson, D. W., & Johnson, F. P. (2009). *Joining together. Group theory and group skills*. Pearson.
Kahneman, D., & Tversky, A. (1982). The psychology of preferences. *Scientific American, 246*(1), 160–173. https://doi.org/10.1038/scientificamerican0182-160
Klöckner, C. A., & Matthies, E. (2004). How habits interfere with norm-directed behaviour: A normative decision-making model for travel mode choice. *Journal of Environmental Psychology, 24*(3), 319–327.
Kullenberg, C., & Kasperowski, D. (2016). What is citizen science? – A Scientometric meta-analysis. *PLoS One, 11*(1), 1–16.
Löfström, E. (2008). Visualizing energy in households: the De-domestication of socio-technical systems and individual- as well as artefact-bound energy use (PhD thesis). Linköping University, Sweden.
Löfström, E. (2014a). Smart meters and people using the grid: Exploring the potential benefits of AMR-technology. *Energy Procedia, 58*, 65–72.
Löfström, E. (2014b). Supporting low-emission lifestyles: Involving future residents at Brøset. In S. Larssæther (Ed.), *Utopia revisited. Towards a carbon-neutral neighbourhood at Brøset*. Fagbokforlaget.
Löfström, E., & Peterssen, I. N. (2011). Public and private feedback: Food-related eco-Visualisations promoting visible symbolic qualities of consumption [Paper presentation]. ECEEE 2011.
Löfström, E., & Svanæs, D. (2017). Eco-visualization: An exploration of the concept and its practical implications. In *Proceedings of the ECEEE Summer Study Proceedings of Toulon/Hyéres*. Stockholm.
Löfström, E., Klöckner, C. A., & Nesvold, I. H. (2020). Nature in your face – Disruptive climate change communication and eco-visualization as part of a garden-based learning approach involving primary school children and teachers in co-creating the future. *Frontiers in Psychology*, 11, 1–10.
Lorek, S., & Vergragt, P. J. (2015). Sustainable consumption as a systemic challenge: Inter- and transdisciplinary research and research questions. In L. A. Reisch & J. Thøgersen (Eds.), *Handbook of research on sustainable consumption* (pp. 19–32). Edward Elgar Publishing.
Morris, R. (2016). *The fundamentals of product design*. Bloomsbury Publishing.
Nachmanovitch, S. (1990). *Free play: Improvisation in life and art*. Penguin-Putnam.
Palensky, P., & Kupzog, F. (2013). Smart grids. *Annual Review of Environment and Resources, 38*, 201–226. (Volume publication date October 2013).
Russell, S. V., Young, C. W., Unsworth, K. L., & Robinson, C. (2017). Bringing habits and emotions into food waste behaviour. *Resources, Conservation and Recycling, 125*, 107–114. https://doi.org/10.1016/j.resconrec.2017.06.007
Samuelson, W., & Zeckhauser, R. (1988). Status quo bias in decision making. *Journal of Risk and Uncertainty, 1*, 7–59.
Schuurman, G. W., Cole, D. N., Cravens, A. E., Covington, S., Crausbay, S. D., Hoffman, C. H., Lawrence, D. J., Magness, D. R., Morton, J. M., Nelson, E. A., & O'Malley, R. (2022). Navigating ecological transformation: Resist–accept–direct as a path to a new resource management paradigm. *BioScience, 72*(1), 16–29.

Stets, J. E., & Burke, P. J. (2000). Identity theory and social identity theory. *Social Psychology Quarterly, 63*(3), 224–237.

Verplanken, B., Aarts, H., van Knippenberg, A., & van Knippenberg, C. (1994). Attitude versus general habit: Antecedents of travel mode choice. *Journal of Applied Social Psychology, 24*(4), 285–300.

Weber, A. (2021). Mobile apps as a sustainable shopping guide: The effect of eco-score rankings on sustainable food choice. *Appetite, 167*, 2021.

Yang, M. C. (2005). A study of prototypes, design activity, and design outcome. *Design Studies, 26*(6), 649–669.

Chapter 7
Global Disruptive Communication: The Thin Line Between Destruction and Disruption in Intercultural Research

Isabel Richter, Maja Grünzner, and Christian A. Klöckner

7.1 Introduction

Culture dictates how we behave and communicate, what is acceptable, and what is not. This is also the core of Kaya's story. Kaya is a fictional, young South African broadcast journalist that we are introducing to illustrate the experiences people are likely to make when they are leaving their cultural comfort zone.

In this short story, Kaya is visiting Japan to contribute to a documentary.

> I arrived in Osaka, Japan to join the South African team of Japan Television International. We were sent off to travel inside Japan to conduct in-depth reports. This is something that my colleagues and I enjoyed a lot. We experienced different landscapes, traditions and environments whilst travelling between different provinces.
>
> The little anecdote I'd like to talk about happened during a three-day trip in the Kansai region. Upon our arrival at our host's house, we were treated to lunch including Soba - the famous hand-pulled buckwheat noodles usually served in Ramen, a pork and miso-flavoured soup. Ramen is a common meal across Japan, mostly served in small roadside restaurants. This dish has been a staple diet for thousands of years. My ordeal began when the host and the other local guests started eating their noodles. I suddenly found myself in the middle of a symphony of slurping sounds. I had no idea how to react. Seeing my surprise, the person sitting next to me whispered that I had to join the slurping in order not to appear impolite and upset the host.
>
> Impolite? My parents in South Africa would have told me off if I would have ever slurped Bobotie, the traditional South African dish made from minced meat, spices and fruit. I started sweating, fearing that I would embarrass myself or worse – not be able to slurp well enough to make our hosts happy. I grabbed the Soba with my chopsticks, took a big breath and tried to delight my colleagues and hosts by joining the slurping chorus. What a mess I caused! A fountain of broth sprayed over the table making me drop half the noodles out of my mouth and back into the bowl. The room went quiet instantly. Multiple pairs of

---

The authors "Isabel Richter, Maja Grünzner & Christian A. Klöckner" are supported by "Sergio Byron Morera Julca & Carya Maharja".

C. A. Klöckner, E. Löfström, *Disruptive Environmental Communication*, Psychology and Our Planet, https://doi.org/10.1007/978-3-031-17165-9_7

eyes were resting on me in a mixture of curiosity and disgust. 'You need to suck harder', the person next to me said quietly but encouraging. I tried. Oil and Ramen splattered over my shirt, like wet paint being flicked onto a canvas. The soba felt thick and heavy and I was confused about how everyone else managed to slurp the noodles without accidents. I gave up trying and asked the locals for help. After bursting into loud laughter, the locals were eager to explain to me the correct technique. The trick was to use the chopsticks to arrange the noodles delicately to capture the right amount of broth, and at the right angle to eventually reach my mouth successfully. I felt proud when I finally managed an almost perfect slurp. Slurping is harder than it seems; however, I learned to appreciate this novel experience throughout my stay and looking back, this is what I remember, and miss, the most.

Ramen, soba, and other Japanese Noodle dishes are fast food; the slurping helps to eat quickly. This prevents the noodles absorbing the broth from getting cold. It is also perceived as a bad manner if excess noodles are bitten off and fall back into the bowl, underlining the importance of mastering the slurping technique properly. Food scientists argue that slurping increases the taste experience with the inclusion of olfactory receptors, which means that you experience the smell of your food through the cavity that connects your mouth and nose (Youssef et al., 2017). Interestingly, the researchers found that people's cultural background determines if slurping or sipping, as an alternative, is creating a better perception of taste.

Just as the story of Kaya shows, visiting another culture can be a disruptive experience. Some customs can appear strange, shocking, or off-putting. What is perceived as normal in one cultural context may be (highly) disruptive in another. It is useful to break out of our established ways of thinking, keep an open mind, and try to understand rather than judge foreign habits to use the power of disruption as a catalyst for positive change.

For disruptive communication, it is key to understand what different cultures perceive as disrupting. In this chapter, we will discuss disruptive communication across cultures.[1] In this chapter, our special focus will lie on so-called non-WEIRD (Western, educated, industrialized, rich, and democratic) populations and people living in developing countries. Even though people living in developing countries are the ones most affected by climate change effects such as extreme weather events or sea-level rise, they are being largely excluded from the global climate discourse (Biermann & Möller, 2019). We will explore how disruptive communication elements can look like in a non-WEIRD population and how they can be co-created. Further, we will explore the limits of disruptive communication, as distinguished from destructive communication, across cultures. Finally, we will discuss how conducting research in non-WEIRD countries by WEIRD researchers (and vice versa) can be a disruption by itself and how we can make sure to distinguish beneficial from adverse effects.

[1] Culture is a complex concept with various definitions. Triandis (2001, p. 3) describes a subjective culture in his work as "the human-made part of the environment." Language and social systems (e.g., economic, educational, political, religious, etc.) are elements of culture. It consists of shared patterns of beliefs, attitudes, norms, role perception, and values as well as consisting of shared practices among a group (Triandis, 2001). Moreover, it is important to point out that it is not given that all people from the same country share the same cultural patterns.

7.2 Non-WEIRD Populations

Looking at the published literature in social and environmental psychology (and beyond), it becomes apparent that samples coming from Western, educated, industrialized, rich, and democratic (WEIRD) societies are overrepresented (Arnett, 2016). These samples represent the country of residence of the authors of the studies, in most cases consisting of undergraduate students from the United States, Canada, or Western Europe. Still, WEIRD people only represent 12% of the world population and are therefore by no means representative of humanity as a whole (Henrich et al., 2010). Instead, almost 60% of the world population lives in Asia, another 20% in Africa with the majority having educational levels of less than secondary school. Despite some progress being made, education, wealth, and opportunities are increasingly concentrated at the top and two-thirds of the global populations are living in countries where inequality follows a growing trend (United Nations, 2020). Less than 12% of humanity live in an industrialized country, less than 5% in a full democracy (Bates, 2020). Reasons for the sampling bias in research are, for example, the unequal distribution of funding and opportunities, the main language in international research being English, and the fact that researchers from developing countries often absolve their higher education in developed countries (Nielsen et al., 2017). Despite a small but significant drop of WEIRD sampling from 95% to 92% in recent years which is mainly due to more publications in psychology is coming from Asian countries, the representation of English-speaking nations is still overwhelming (Newson et al., 2020; Rad et al., 2018). What does this bias mean for the sustainability discussion, and how can we work toward more decolonized research, potentially featuring disruptive communication in the future?

Most studies published in leading psychology journals rely on WEIRD samples as a matter of course but then use their findings to make general conclusions about all humanity (Cheon et al., 2020; DeJesus et al., 2019; Hendriks et al., 2019; Rad et al., 2018; Tam & Milfont, 2020). This poses the question of how legitimate the conclusions and inferences are about thoughts, feelings, communication, and behavior of humans in general that are drawn from nonrepresentative, WEIRD populations. Bates (2020) claims that WEIRD samples are not a justifiable testing ground for generalized theories concerning communicative behaviors in friendship, relationships, politics, or health. Generalizing from one population to another stands in strong conflict with one of the top recommendations for effective communication, which is to know your audience and to adapt your strategy of communication depending on who you are talking to (Markowitz & Guckian, 2018; Moser, 2014; Richards & Carruthers Den Hoed, 2018). According to principles of intervention design for behavior change, we need to identify people's behavioral drivers and barriers (Klöckner, 2013; Steg & Vlek, 2009), which may be different in different countries, cultures, and contexts. Using disruptive communication to make people stop and think (see Chap. 1) might therefore need different kinds of disruptors, depending on who we want to address.

7.3 Similarities and Differences Between Cultures

To give some examples, all humans have in common, how they express the five basic emotions (anger, disgust, fear, sadness, and surprise) via facial and bodily expressions (Ekman et al., 1969; Witkower & Tracy, 2020). People differ, however, to what extent they experience these emotions and how culturally accepted they are. Further examples for commonalities are the way we perceive color, extract numerical information, or develop social relationships (for an overview see Henrich et al., 2010). Key differences between populations can be found in risk perception and aversion (Henrich & McElreath, 2002), future discounting (Godoy et al., 2004), punishment and cooperation (Herrmann et al., 2008), moral reasoning (Haidt et al., 1993), self-concept (Heine, 2008; Markus & Kitayama, 2010), values (Schwartz, 1992, 2012), and even basic sensations such as visual perception (for an overview see Henrich et al., 2010).

By acknowledging the similarities and differences between populations, we can adapt our communication strategies, not only to make them more efficient but also to avoid interventions to become potentially destructive. It is therefore paramount to get to know your audience before developing and implementing a novel communication approach.

Revisiting the signs of planetary distress outlined in Chap. 1, such as droughts, sea-level rise, storms, and extreme weather events, we find that the majority of these events affect people in developing countries the most (Eckstein et al., 2019). They are, however, mainly caused by industrialized nation's lifestyles. The solutions outlined in Chap. 1, such as insulation of large houses, electrification of car traffic, and reducing overconsumption, are solutions to make lifestyles in the Global North less environmentally damaging. Disruptive communication has been focused on behaviors (e.g., following a vegetarian diet) and their drivers in WEIRD populations (e.g., Zur & Klöckner, 2014), openly questioning modern lifestyles and the damage they do to the planet. In non-WEIRD populations, we most likely need to adopt a very different approach, address different behaviors, and consider different pressures, circumstances, and lifestyles. As thoroughly explained in Chap. 1, there is a large array of gentle approaches such as nudging, incentivizing, or campaigning that are supposed to convince citizens of more sustainable behaviors without upsetting them. The concept of disruptive communication calls for an unrulier approach, risking upsetting or shocking people along the way.

Is this disruptive approach equally appropriate for sustainable change-making in developing countries, and if so, what do we need to consider? In the next section, we elaborate on psychological drivers of environmental relevant behaviors in non-WEIRD as opposed to WEIRD populations. This helps us to gain a deeper understanding of why generalizing from WEIRD samples to the rest of humanity might not be enough.

7.4 Behavioral Predictors Around the World

Phelan et al. (2020) explored the mental models of plastic pollution in remote island communities in Sulawesi, Indonesia. In contrast to their expectations, they found that personal knowledge or awareness as well as the concept of individual behavior change and factors a large body of environmental psychology literature revolves around are absent from the mental models of the island populations. Instead, the mental models' centerpiece consisted of collective and social factors such as community-organized waste disposal and collective beach cleans. Social factors might be more important than individual factors in small, subsistence-oriented communities. In line with this thought, Truelove et al. (2015) found that the identification with their home community was the strongest predictor for climate change adaptation behavior in a sample of paddy farmers in Sri Lanka.

Let us experiment and look at the predictors of one behavior, such as household waste management, across samples. In their study looking at predictors of household waste management in Indonesia, Simmons and Fielding (2019) identify perceptions of positive change in their community and specific attitudes as the main behavioral drivers. Factors such as descriptive norms that have been important in a large number of studies conducted with WEIRD samples played a minor role in the Indonesian fishing communities.

Along the same lines, external circumstances such as an existing recycling scheme was the most relevant factor (as compared to attitudes or demographics) explaining if people engage in household waste management in a study by Singhirunnusorn et al. (2012) conducted in Thailand. In Bangkok, a community-based recycling project distributing eggs to people delivering their recyclables successfully influenced household waste management (Kladnuenklum, 2003). In Iran, moral obligation to recycle has been found to be the strongest predictor for household waste management (Pakpour et al., 2014), accentuating the importance of morality in the Iranian culture as a springboard for communication in this specific context.

We can see that not only behaviors and their drivers differ between WEIRD and non-WEIRD populations, but they also differ between countries and regions. Understanding the population we engage with, their behavior and its drivers should therefore be paramount and always our first step before we jump into intervention development.

The Story of Bu Mahariah

Bu Mahariah is a preacher from the island of Panggang, Kepulauan Seribu, Indonesia. Her ways of campaigning against plastic waste are disruptive for many of her community members.

Bu Mahariah is applying her experience from her job as a religious teacher and preacher in local Quran reading groups to inspire more sustainable handling of household waste and plastic litter. By connecting both,

(continued)

sustainable and religious values into her sermon, she inspires her community to engage in exemplary waste management. One of her initiatives is the collection and transformation of domestic waste into compost which is then distributed to the local communities to grow food. Her (mostly female) community group also performs regular beach cleans and uses the collected plastic waste as material for craftworks. These are then sold by her group members to provide additional income to the community as can be seen in this short film by the Indonesian filmmaker Carya Mahajra https://www.youtube.com/watch?v=L607V4dyjwA.

Bu Mahariah illustrates the importance of communicating in a way that speaks to the values and emotions of the audience. Plastic pollution is not a topic that is typically raised within religious conversations. Instead, religiosity and rationality are perceived as conflicting themes on the island. Using religion to drive the sustainability agenda is a daring approach, potentially surprising many community members at first. In addition, Bu Mahariah stands out as a female leader in her conservative, patriarchal Indonesian society. Especially at the beginning of her mission, she was confronted with a lot of resistance, and people reacted with some degree of rejection and reluctance.

This example shows how a moment of surprise or novelty, paired with tailor-made communication and co-developed solutions can lead to positive change, in line with the approach proposed in this book.

For effective communication strategies, it is relevant not only to identify behavioral drivers but also behavioral barriers. The widely discussed "29 dragons of climate inaction" illustrate psychological obstacles to adequate climate change adaptation and mitigation (Gifford, 2011). These dragons are, however, mainly based on research coming from Western samples and formulated to fit current Western culture (e.g., the dragon of techno-salvation or environmental numbness). While some of the dragons surely are universal and reflect obstacles WEIRD and non-WEIRD populations face alike, it seems like some dragons are still missing. The dragon of "basic struggle for survival," "temporal discounting," or "continuation of traditional practices" might be three additional species we might want to add to the dragon typology and thereby make the dragon family more inclusive. The very fact that *dragons* represent obstacles and blockages is a Western perspective on these magic animals compared to the benevolent and luck-bringing dragons in Asian cultures as Gifford (2011) mentions in a footnote.

We conclude that despite classifications like Gifford's dragons of inaction being highly useful to identify psychological barriers to sustainable behavior, we need to critically question if these models apply to WEIRD and non-WEIRD populations alike. For future research, we would like to suggest opening up our existing models for inclusive adaptation, to make them suitable for global collaborations.

7.5 Disruptive Communication Without Being Destructive

To convince people to change their behavior, we would wish that the same style of communication deems successful around the world; that disruptive communication addressing certain environmental issues such as climate change could be applied the same way across diverse contexts. Sometimes this works: the Friday for Future movement, for example, motivates humans, especially the youth, all around the world to speak up and protest for drastic changes in our society (see Chaps. 1 and 4). However, the situation-specific challenges differ significantly between countries and cultures. We, therefore, suggest considering the three elements of the Culture-Person-Situation model (Fig. 7.1; Bosse, 2010; Leenen et al., 2005) when attempting disruptive communication in intercultural settings. The model is widely used in intercultural competence training and argues that communication in an intercultural setting will only be successful if we take into account the interplay of cultural, personal, and situational factors (Bosse, 2010; Leenen et al., 2005). This Culture-Person-Situation model can be used as a framework to identify behavioral drivers and barriers as well as to evaluate disruptive communications.

Disruptive communication is provocative by nature. Instead of gently pointing toward environmental problems, disruptive communication aims to position them right in front of us, literally "In our Face" (Löfström et al., 2020). This is an innovative and daring approach already in WEIRD populations. In non-WEIRD populations, it is important to consider that not only the problems and behaviors are considerably different from the Global North but that people also often struggle with basic survival. We need to ask ourselves the question if it is ethical to use a disruptive approach and identify the thin line between disruptive and destructive. If disruptive communication aims to interrupt citizens in their "standard way of behaving" (see Chap. 1), we need to be aware that while this might be an annoyance in the WEIRD populations, it can lead to severe consequences in non-WEIRD populations. The example used in Chap. 1, an artistic performance theatre blocking the street and keeping you from driving your usual route to your workplace will most likely be an inconvenience in industrialized countries. If we interrupt a fisherman in a developing country from going fishing for one day, his family might go hungry, and he might lose crucial income for basic survival. To effectively communicate with communities around the world, we should not only be aware of behavioral drivers and barriers as described in the previous section but also think about potential consequences. We suggest that the best way to distinguish disruptive from destructive is, again, specifically tailoring your communication strategy to your audience and the audience's circumstances as the Culture-Person-Situation model suggests.

In a study by Richter et al. (n.d.-b), the authors evaluated different ways of communicating to coastal communities in the Philippines. The group found that the most effective messages had an emotional and a behavioral dimension. More important though was the relevance of the message to the target audience. To explain this with an example, high school students ranked messages about seaweed farming as

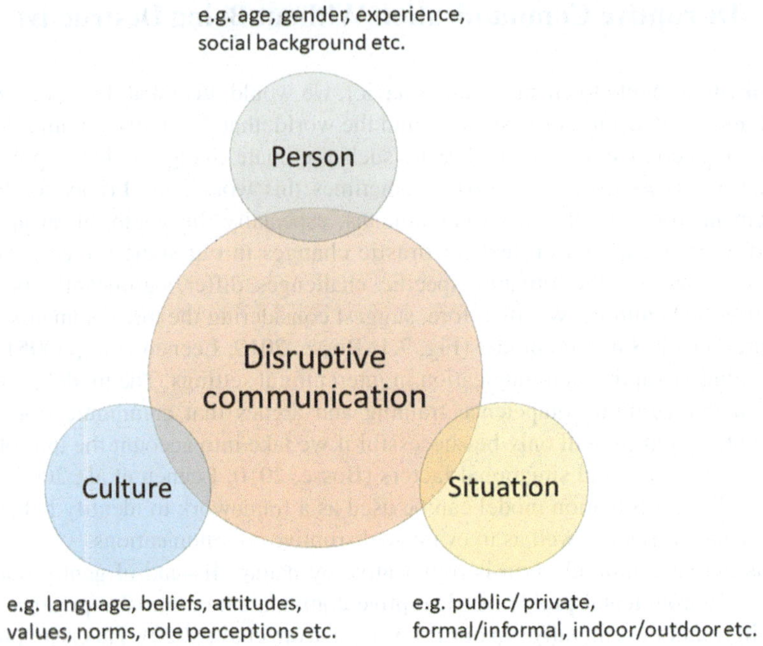

Fig. 7.1 Disruptive communication in the light of the Culture-Person-Situation model. (Adapted from Bosse, 2010; Leenen et al., 2005)

very boring and ineffective. Fisherfolk, especially from an island where fishing grounds were almost depleted, ranked the same message as very important and effective. The researchers concluded that effective messages do not depend much on the layout of the message delivered. Instead, the applicability of the message content to the audience is key. This research points toward a medley of emotions, solutions, and applicability as being the recipe for successful communication.

Successful disruptive communication depends on similar criteria (see Chap. 1): with the disruption, people shall be involved emotionally and make room for engagement ("stop and think"). As a next step, these emotions become a catalyst for solution development before being communicated to relevant audiences. For disruptive communication to become effective in the non-WEIRD populations and to avoid being destructive, we suggest adding the dimension of *applicability*. Only if disruptive communication is channelled into applicable solutions, it combines added value for both, the audience and the environment.

Tales from the Field #1: Palawan, the Philippines

As part of the GCRF Blue Communities project (https://www.blue-communities.org/Home), I spent some time in Palawan, an island in the southwest of the Philippine Archipelago and UNESCO Man and Biosphere Reserve. These weeks that I spent together with our local partners have been a positively disruptive experience for me. The collaboration taught me a lot about the significance of building relationships, co-creation and adaptation, and why it is so important to take the back seat and listen when you arrive in a culture that is not yours.

Before embarking to the Philippines, my group and I made exact activity plans for every single day of my stay. Surveys and PowerPoint slides had been prepared, lectures had been written, and workshop participants had been invited. However, things did not quite turn out as expected.

As a German, I almost obsess over efficiency. According to my schedule, our activities were planned to start at 8 am after I arrived on the island. I knew the journey would be long and the temperatures would be hot, but I completely underestimated how hard it is to ignore heavy jetlag. The result was that I fell into a deep sleep in my room and missed the whole first day. The first lesson was learned.

Further, I underestimated the role of mealtimes. As a Westerner, I suspected a squeezed-in, sandwich-type lunch and not a 2-hour break with a colorful banquet every day. For the local partners, this was the time not only to recharge but also to discuss ideas, consult experts, or obtain materials. In the beginning, I perceived the long break as distressing, and the feeling of not being efficient was haunting me. After a while, I learned to appreciate the creative time and used it for a walk, a conversation, or simply for a nap. Overall, I discovered napping as a wonderful new way of recovery, something that I never thought of doing in my home environment.

Another way in which my rigid mindset was challenged was the constant adaptation and change of plans I had to accept. Every day, plans had to be readjusted due to participants not turning up, people joining who have not been expected, or a heavy rainstorm. The day I remember the most was the day of a big data collection involving 150 university students in Puerto Princesa, the capital city of Palawan. The designated lecture hall for the activity was ideal: it had a huge projector screen, enough seating space for everyone, and air-conditioning, an asset during the dry season with temperatures reaching 40 °C. When a sudden electricity blackout across the whole island disrupted the activity, I expected that we would have to call everything off. My colleagues, however, immediately reacted, and the whole group was moved to another room within minutes. The new room was small, did not allow everyone to sit, had a small screen, and had no air condition. Somehow, I seemed to be the only one concerned. Everyone was motivated to continue; the data collection was perfectly undertaken, even when the temperatures crept up to 50 °C. I watched with unbelieving amazement how my colleagues remained calm, positive, and entertaining throughout.

This experience has been truly disruptive for me. I had to reassess the value of key qualities such as flexibility and optimism. But my colleagues did not only teach me serenity; I brought home a lot of knowledge about how to dress appropriately in stakeholder meetings, how to communicate with local fishermen, how to negotiate the correct price for a tricycle, or how to snap crabs with my bare hands.

I feel humbled when I think back to my stay in the Philippines and how it taught me that I still have so much to learn. I am convinced that I am not the only one making this erroneous assumption that our (research) approaches are universal. Herby, I would like to invite our readers to question their established worldviews and to look out for opportunities to "stop and think."

—Isabel Richter

7.6 Scenarios as a Disruptive Communication Tool

A major challenge in the sustainability discussion is the mismatch between short-term personal, political and economic benefits, and long-term outcomes related to global sustainability (Fauré, 2016). The psychological, geographical, and temporal distance between today's behavior and its future consequences often impedes climate action (Spence et al., 2012). People develop cognitive biases of future thinking depending on their economic, environmental, educational, and social environment and upbringing (Ellis et al., 2017). Mentally simulating possible futures in as much detail as possible can make the future seem "closer than we think" as suggested by Wang et al. (2019). Through mental simulation, future scenarios can potentially bridge the psychological distance between "now and then" and overcome perceptual biases of future thinking (for an overview see Richter et al., n.d.-a). Mental simulations can also evoke fear, hope, and other emotions, which are suspected to have a stronger influence on behavior than do cognitive evaluations (Loewenstein et al., 2001). Given their characteristics, future scenarios can serve as a form of disruptive communication.

People in marginalized communities discount the future ten times more than WEIRD people (Godoy et al., 2004). Living in an unpredictable environment shaped by ecological, social, and economic uncertainty elevates the preference for short-term benefits and orientation to the present rather than the future (Gurven, 2018). Present-oriented societies are more likely to have lower rates of preventive behaviors such as exercise or nutritious diets and engage more in risky behaviors such as early sexual activity, alcoholism, or drug abuse (Chisholm et al., 1993; Huston & Finke, 2003; Kosteas, 2015). Across 46 nations, people living in better and safer conditions are more likely to wait for a delayed reward as compared to an immediate reward (Bulley & Pepper, 2017). We need to be aware of these differences when we direct our communication toward sustainability, which is by nature a long-term future-directed concept.

An example of co-developed future scenarios as disruptive communication is the board game Savanna Life (http://www.savannalife.no/ – see also Chap. 5 for a discussion of games as disruptive communication elements). Graae and her team aimed to form a common ground for discussing relevant issues around sustainable development between local communities living close to protected areas and other stakeholders (Savanna Life Project, 2019). The game was developed over numerous iterations by a team consisting of international, interdisciplinary researchers, an experienced game designer, as well as local stakeholders "as a learning tool to enable local households and communities to evaluate the consequences of livelihood strategy choices and investment, and to safely explore alternative strategies" (Saíd et al., 2019, p. 76). In line with the disruption stage of disruptive communication, Savanna Life allows players to take risks and experience dangerous situations outside their control. Naturally, the game provides the frame in which players are moving within ("framing" stage) as well as providing them with various behavioral options to influence the outcome ("twisting" stage) – all in a safe environment. Building on the game experience, facilitators can help to encourage the game players' reflection on the experienced scenarios and draw connections to the real world ("using" stage; see more in Chap. 5). The role of relationship building and respecting local customs is described in the Tales from the field #2 below.

Another example of future scenarios as a potentially disruptive communication tool comes again from the Philippines and the Blue Communities project. Together with a variety of local stakeholders, three future scenarios have been co-developed and subsequently evaluated for their effects. One of these scenarios reflected the "expected" future, business as usual, whereas the other two scenarios venture off toward the unexpected: one apocalyptic scenario (worst case) and one utopian scenario (best case). The rationale of disruptive communication would suggest that the unexpected scenarios cause stronger reactions as compared to the expected scenario. This was confirmed by the researcher's findings (Richter et al., n.d.-a, 2021). When the scenarios are presented to different audiences, the unexpected scenarios are superior to the expected scenarios in promoting intentions for sustainable behavior change. The utopian scenario, for example, significantly increases intentions to engage in sustainable community activities and evokes strong positive emotions such as hope. The apocalyptic scenario leads to strong negative emotions such as fear or anger. The researchers further find that people consider future consequences more after creatively engaging with future scenarios. Relating these findings to disruptive communication, we find that scenarios holding a dimension of unexpectedness seem to evoke strong emotions and thereby disrupt people's established ways

Fig. 7.2 Co-created mural paintings illustrating future scenarios of the administrative region of Taytay in Palawan, Philippines

of thinking and acting. Channelling this moment of disruption into the development of positive, shared visions can potentially inspire sustainable behavior change.

Further ways of potentially disruptive, inclusive, and co-created communication that have been appreciated within developing communities are, for example, photovoice (Lim et al., 2020), storytelling (Iseke, 2013), theatre plays (Ansloos & Wager, 2020), or co-created paintings as shown in Fig. 7.2 (Richter et al., 2021). It is, however, population- and context-specific, which way of communication is most appropriate.

Tales from the Field #2: Maasai Mara National Reserve, Kenya

I was excited and nervous when I got the opportunity to become part of the interdisciplinary and multicultural team of the Savanna Life project (http://www.savannalife.no) evaluating a board game as a communication tool in Kenya.

I have always been fascinated by intercultural similarities and differences. During my masters, I was able to learn about the Psychology of Intercultural Relations. I will never forget the endless discussion with my peers and lecturers about why we think and behave the way we do. These lectures and conversations alongside adapting to live in a country (Portugal) where I did not even speak the language made me understand that some things that I thought as

(continued)

being normal are not perceived as normal by others – and vice versa. Some gestures (kisses on the cheek as hello from my Portuguese peers) or keeping a personal distance (out of respect to my Chinese peers) made me question my behaviors and their universality. Being in such a multicultural environment allowed me to develop openness and embracing each other's differences preparing me for my first multicultural research project, Savanna Life.

The core tool of the project was the board game called "Savanna Life." The game follows the narrative of a community in East Africa and represents four families – each player embodying one of them. The goal of the board game is to bring wealth to your community and family by making sustainable choices. Moreover, the game provides unforeseen challenges during the play in form of disruptive events – negative and positive – such as wildlife attacks or foreign stakeholders investing in new technologies. Even though players cannot control the events that are going to happen, they have control over how they collaboratively handle them. The game provides a safe and fun environment where each player can step outside their everyday life and try out various behaviors which ideally creates a state of reflection (see Chap. 5.2).

The project's success was based on the years of experience and network development of the Savanna Life project lead and her team. They established close relationships with local researchers from Kenya and diverse local stakeholders from communities around the Serengeti-Mara Ecosystem, who took on the role of organizers and community facilitators. Having team members who know the communities and the local customs and who speak the local languages fluently (in this case Swahili and Maa) helped to promote clarity and avoid misunderstandings.

To adapt to local customs, it was, for example, agreed to start each activity with a shared prayer to set the scene, something that would be unusual to exclusively WEIRD research settings. Further, female and male players were separated to give them the possibility to reflect on the game. This was based on a previous experience where female players appeared to hold back their conversations when male participants were present.

A disruptive experience for us visitors was when on one day our research activity had to be arranged in a very small hut. It only had one small window and a bench inside – no chairs, no tables. I was worried about how we would make everyone fit into the small hut and if participants would be upset about the arrangement. Even though it felt disrupting to me, I also felt inspired by the spirit of my collaborators and decided to focus on what seemed most important: creating a pleasant environment for everyone. Eventually, we set up two board games on the floor (see Fig. 7.3).

Until today, I am amazed about the outcome: people sitting on the floor, laughing, and enjoying the board game experience.

The work with Savanna Life taught me important lessons for research in, to me, foreign cultures.

Intercultural research experiences can be successful if we stay open-minded, welcome disruption, create a shared goal, trust the value of communication, and have fun!

—*Maja Grünzner*

Fig. 7.3 Community members and researchers engaging in the Savanna Life Board game close to the Maasai Mara National Reserve, Kenya

Summing up this section, thinking and planning far ahead is counterintuitive for many people, especially for those experiencing economic hardship. Co-creating future scenarios might be a way to empower people to question their established ways and imagine different possible futures. Especially when these scenarios involve a dimension of unexpectedness, they act as a disruption and, eventually, a catalyst for behavior change.

7.7 Intercultural Research as a Disruption

So far, we have been talking about WEIRD and non-WEIRD samples, differences, and similarities between populations and communication techniques and tools. In this section, we would like to mention a few important considerations regarding the research process itself. These considerations will help not to accidentally offend collaborators and to facilitate co-creation and relationship building. We will further discuss the opportunity of intentionally cross-cultural norms to use the research process as a positive disruption by itself.

In their article, Simonds and Christopher (2013) explain how during their collaboration with the Native American Crow tribe they encountered several critical situations and which lessons they learned. Unknowingly, they crossed cultural boundaries during the research process. As researchers coming from a Western tradition, they were trained to break data apart, anonymize it, and systematically model relationships. To their Native American collaborators who use storytelling as their main form of communication and for whom breaking stories apart is disrespectful to the narrator, this way of structuring and deducing has been perceived as an offence. The tribe, therefore, decided not to show interest in the project anymore and started to be quiet and absent. This collaboration has been perceived as destructive by the members of the Crow tribe. Only through dialogue, careful adaptation of the Western research methods and the development of mutual trust, the partners could find a way to turn this disruption into a productive partnership again.

A reflection we should make right at the start of any cross-cultural research interaction is how we might appear to our colleagues and how a more Western self-concept might be perceived within non-Western populations. An overly heightened self-presentation might be perceived as uncomfortable, rude, or implausible and pose a barrier to the development of a trusting relationship. As an example, Kim (2002) starts her chapter on cross-cultural perceptions of bragging with two popular quotes, one from a WEIRD and from a non-WEIRD author, clearly showing the stark contrast of how socially accepted self-presentation looks like: "Mature rice plants bow down lower" is a Korean expression indicating humble self-presentation. "Unless you blow your own horn, who will?" represents the American counterpart. These very contradicting approaches to self-presentation provide a typical moment for involuntary disruption before even starting the research process.

Being aware of such cultural conventions before interacting with our partners will help to avoid creating unintentional, negative (first) impressions. Browne-Nuñez and Jonker (2008) found in their review that perceived social pressure might also result in responses of participants being biased toward what they "think the researchers want to hear" (p. 64). Participants might assume that the researchers are representatives of certain interest groups such as wildlife authorities. In their responses, they will then avoid mentioning taboo subjects such as intergroup conflicts or poaching (Browne-Nuñez & Jonker, 2008). Suspicion can arise when foreign researchers demonstrate unfamiliar or inappropriate behavior or do not accustom to normative appearance, clothing, or gender roles. Even if this interviewer bias in cross-cultural research can never be eliminated, it can potentially be reduced through specific training sessions with the help of local community facilitators.

Standard (WEIRD) psychology research methods strive for randomization and representativeness to achieve valid and generalizable data. When working with non-WEIRD samples, however, sample size and composition are often determined by the availability of participants (e.g., villagers in rural areas) as has been the case in several studies in Africa (Browne-Nuñez & Jonker, 2008). People living in subsistence communities have to weigh up the pros and cons of participation in a scientific study carefully as it means potentially losing one day of income generated by fishing or hunting. This often results in biased or unbalanced samples, as was the case, for example, in Sumeldan et al. (2021). In their data collection in Palawan, Philippines, the sample consisted almost exclusively of women as they were staying at home with their children while their husbands went out fishing or farming and thereby acted as willing participants for the project interviews.

Another seeming sampling bias might lie in the systematic in- or exclusion of some subgroups. Local informants or collaborators might, willingly or unwillingly, point the researchers toward specific people or groups and not toward others, which creates a systematic sampling bias. Stratified sampling can lead to conflicts, especially if some subgroups are disregarded by the majority of the population. As an example, the population living in Tun Mustapha Park, Northern Sabah, Malaysia consists to some extent of undocumented, stateless people who live in floating villages off the coast. These foreign workers are often demonized as causing social, economic, and security concerns and are not acknowledged by the government as part of the Sabahan population, which makes them an "invisible" subgroup that is hardly involved in any scientific activity (Allerton, 2017; Hasbullah, 2020; Warsono et al., 2020).

We have the option to intentionally involve discriminated subgroups into research to create a disruptive element. The crucial element here is *intentionality*. If we want to disrupt with novel approaches and uncommon samples, we should do this in an intentional, informed way, supported by local experts and thereby prevent destructive outcomes. The key to distinguish between disruptive and destructive is co-developing the research process as well as its communication and dissemination. This might mean the appointment of project-internal expert groups for certain parts of the project. Members of the local communities are appointed experts of local knowledge for example, whereas the international researcher might become the expert for his or her competence. Like this, various forms of knowledge will be appreciated and both, cultural appropriateness and scientific rigor, are accomplished. When results are obtained, we have the option to not only share our findings with our academic peers but also with respective populations. This might become another form of disruptive communication because the findings might be painful or disturbing. Sharing research insights with the community understandably and appropriately, as lined out in precious sections, can, however, enrich the dialogue between researchers and locals and clear the path for sustainable solution development. It also opens the opportunity to possibly find new connections, similarities, and differences between the local knowledge and scientific data. As Huntington (2011, p. 183) expresses: "At its heart, the engagement of scientists with traditional knowledge is really about the engagement of people, sharing information about what matters most to them."

Tales from the Field #3: Peruvian Andes

As part of my work as a scientific researcher at the Instituto Geofisico del Peru, I have participated in several national and international projects to improve access to clean water. Here, I will talk about my experience in the "Páramos" and "Eros-IsoGlas" projects carried out in the north and center of Peru. My experiences are an example of how collaborations between researchers, policymakers, and communities can become a disruption by itself and how to harvest positive rather than negative outcomes.

Monitoring water, soil and sediments have complexities and physical demands that can be overcome with training or perseverance. What is harder to overcome is communication barriers between scientist, policymakers, and communities and the resulting lack of acceptance of research projects within the rural population.

In remote places of Peru, such as mountain and forest regions, the majority of the population holds low levels of formal education, making them susceptible to their leaders' influence as their only source of information. In some cases, this can interfere with acceptance or support for research projects because these leaders aim to satisfy their own private or political interests instead of community welfare or sustainability.

We experienced an example of this in the Ronquillo River Basin, a catchment that borders the city of Cajamarca, my hometown. This catchment accommodates five remote villages, each of them representing different political and environmental standpoints.

We aimed to monitor water and sediment quality and quantity and scheduled a meeting with each rural community to reach an agreement for the scientific activities (Fig. 7.4). Three of the five communities immediately agreed to the monitoring activities and offered their support. The fourth community taught us that our communication strategy was still deficient. 18 community members and their leader joined the meeting. We explained the relevance and prospected proceedings of our project. Instead of an agreement, we were faced with rejection. The comments we received ranged from "We are tired of

Fig. 7.4 Community consultation in the Ronquillo River Basin (**a**) and installation of the "El Ronquillo" observatory (**b**)

(continued)

the government, it does nothing for us" and "What are these studies useful for? We need new buildings" or "If you build us a water reservoir, we invite you to Caldo de Gallina [traditional chicken soup]." We even received indirect threats like "We do not allow you to build a station and install the equipment. If the equipment is lost, we may not know."

This reaction disrupted not only our research process; it also led to angry and heated discussions. No common ground could be found, and the conflict seemed insurmountable. It seemed like disruption almost became destruction. Only through conversations with some "campesinos" [friends] in the catchment, we found that fear was the main driver of the communities' rejection. They advised us to change our communication strategy and demonstrate people's benefits in a locally understandable way. With the help of the three supportive communities and the campesinos, we developed a more inclusive communication approach. Eventually, all villages came together to implement the "El Ronquillo" observatory (Fig. 7.4), which to date is the best-monitored basin in Peru. The initial rejection helped us to rethink the way we approach communities. We now know that building personal relationships and speaking the local language are key.

Without inclusive communication, the disruption through scientists poses an additional destructive force upon the already vulnerable Peruvian mountain communities.

—Sergio Morera

Summing up this section, we conclude that the process of conducting cross-cultural research itself balances on the thin line between disruptiveness and destructiveness. If we do not prioritize co-developing the methodology and co-creating the dissemination, we run the risk of unintentionally crossing cultural norms and thereby being destructive. We need to be open to coaching and take advice from our partners when it comes to self-presentation, personal interactions, language, and sampling. However, if prepared carefully, crossing cultural norms can also be a catalytic disruption, potentially giving an impulse toward sustainable change.

7.8 Conclusions

As discussed at the beginning of this book (see Chap. 1), politics and behavioral sciences are trying to avoid pushing people out of their comfort zone when it comes to sustainable change processes. However, when we start collaborating across countries and cultures, we are automatically pushed out of our comfort zones.

Therefore, we argue that, just as disruptive communication, intercultural work can be upsetting, may cause resistance, and be unconventional (for one or the other cultural group) but also energizes and can be the start of a process. We believe that intercultural exchange can be a positive disruption itself (see Tales from the field stories #1–3) when done in a respectful and open-minded way. Culture clashes can be useful for breaking out of entrenched ways of thinking and make way for equality, dialogue, and sustainability. The key to make cross-cultural disruption a beneficial experience is to build trusting research relationships with local populations.

Practically, if we aim to potentially disrupt current practices and habits to meet common sustainability goals, we need to think about more than just the message itself. Instead, we need to find out more about the population with which we want to collaborate and communicate. Not everything that might be perceived as disruptive in the developed world is equally effective and appropriate in developing countries. Being aware of intercultural similarities and differences between populations and cultures is a precondition for understanding why people behave the way they do and, as a next step, creating the right approach to change these behaviors. The original definition of disruptive communication calls for unexpectedness, emotional involvement, and direction into creative solution development. In addition to these criteria, we also argue that disruptive communication, especially in developing countries but potentially also elsewhere, should be *intentional* (by the disrupter) and *applicable* (to the disrupted). Thorough evaluation of projects following the principle of disruptive communication will shed more light on the essential ingredients of disruptive communication for sustainability around the world.

When we then develop our disruptive communication approach, it is key to consider the three elements of the Culture-Person-Situation model: who are we communicating with? What cultural background does this population hold, and what is the situation we are communicating in? Intercultural collaborations invite us to be creative in our approaches and dip our toes into arts, games, photography, or scenario development. Again, we need to keep in mind that none of these approaches will be a one-fits-all solution, especially not in a global context.

Only by close collaboration and co-creation beyond our discipline and culture, we can develop communication strategies that are the effective disruption we aim for. Just as Newson et al. (2020) recommend, we need to conduct "work that is not just Worldwide, but also In Situ, Local, and Diverse (WILD)." The art does not lie in expressing yourself as elaborate as possible but in communicating what matters in a way that your target audience is inspired by.

References

Allerton, C. (2017). Contested statelessness in Sabah, Malaysia: Irregularity and the politics of recognition. *Journal of Immigrant & Refugee Studies, 15*(3), 250–268.

Ansloos, J. P., & Wager, A. C. (2020). Surviving in the cracks: A qualitative study with indigenous youth on homelessness and applied community theatre. *International Journal of Qualitative Studies in Education, 33*(1), 50–65.

Arnett, J. J. (2016). The neglected 95%: Why American psychology needs to become less American. *American Psychologist, 63*(7), 602–614.

Bates, B. R. (2020). Making communication scholarship less WEIRD. *Southern Communication Journal, 86*, 1–4.

Biermann, F., & Möller, I. (2019). Rich man's solution? Climate engineering discourses and the marginalization of the Global South. *International Environmental Agreements: Politics, Law and Economics, 19*(2), 151–167.

Bosse, E. (2010). Vielfalt erkunden–ein Konzept für interkulturelles Training an Hochschulen. In Schlüsselqualifikation Interkulturelle Kompetenz an Hochschulen (pp. 109–133). VS Verlag für Sozialwissenschaften.

Browne-Nuñez, C., & Jonker, S. A. (2008). Attitudes toward wildlife and conservation across Africa: A review of survey research. *Human Dimensions of Wildlife, 13*(1), 47–70.

Bulley, A., & Pepper, G. V. (2017). Cross-country relationships between life expectancy, intertemporal choice and age at first birth. *Evolution and Human Behaviour, 38*(5), 652–658.

Cheon, B. K., Melani, I., & Hong, Y.-y. (2020). How USA-centric is psychology? An archival study of implicit assumptions of generalizability of findings to human nature based on origins of study samples. *Social Psychological and Personality Science, 11*(7), 928–937.

Chisholm, J. S., Ellison, P. T., Evans, J., Lee, P., Lieberman, L. S., Pavlik, Z., et al. (1993). Death, hope, and sex: Life-history theory and the development of reproductive strategies [and comments and reply]. *Current Anthropology, 34*(1), 1–24.

DeJesus, J. M., Callanan, M. A., Solis, G., & Gelman, S. A. (2019). Generic language in scientific communication. *Proceedings of the National Academy of Sciences, 116*(37), 18370–18377.

Eckstein, D., Künzel, V., Schäfer, L., & Winges, M. (2019). *Global climate risk index 2020.* Retrieved from Bonn: http://www.germanwatch.org/en/cri

Ekman, P., Sorenson, E. R., & Friesen, W. V. (1969). Pan-cultural elements in facial displays of emotion. *Science, 164*(3875), 86–88.

Ellis, B. J., Bianchi, J., Griskevicius, V., & Frankenhuis, W. E. (2017). Beyond risk and protective factors: An adaptation-based approach to resilience. *Perspectives on Psychological Science, 12*(4), 561–587.

Fauré, E. (2016). Sustainability goals combining social and environmental aspects. KTH Royal Institute of Technology,

Gifford, R. (2011). The dragons of inaction: Psychological barriers that limit climate change mitigation and adaptation. *American Psychologist, 66*(4), 290.

Godoy, R., Byron, E., Reyes-García, V., Leonard, W. R., Patel, K., Apaza, L., et al. (2004). Patience in a foraging-horticultural society: A test of competing hypotheses. *Journal of Anthropological Research, 60*(2), 179–202.

Gurven, M. (2018). Broadening horizons: Sample diversity and socioecological theory are essential to the future of psychological science. *Proceedings of the National Academy of Sciences, 115*(45), 11420–11427.

Haidt, J., Koller, S. H., & Dias, M. G. (1993). Affect, culture, and morality, or is it wrong to eat your dog? *Journal of Personality and Social Psychology, 65*(4), 613.

Hasbullah, N. H. (2020). *Public perception of immigration policy: A case study of Filipino illegal immigrants in Semporna*. International Islamic Univeristy, Kuala Lumpur, Malaysia.

Heine, S. J. (2008). *Cultural psychology*. W. W. Norton.

Hendriks, T., Warren, M. A., Schotanus-Dijkstra, M., Hassankhan, A., Graafsma, T., Bohlmeijer, E., & de Jong, J. (2019). How WEIRD are positive psychology interventions? A bibliometric analysis of randomized controlled trials on the science of well-being. *The Journal of Positive Psychology, 14*(4), 489–501.

Henrich, J., & McElreath, R. (2002). Are peasants risk-averse decision makers? *Current Anthropology, 43*(1), 172–181.

Henrich, J., Heine, S. J., & Norenzayan, A. (2010). Most people are not WEIRD. *Nature, 466*(7302), 29–29.

Herrmann, B., Thöni, C., & Gächter, S. (2008). Antisocial punishment across societies. *Science, 319*(5868), 1362–1367.

Huntington, H. P. (2011). The local perspective. *Nature, 478*(7368), 182–183.

Huston, S. J., & Finke, M. S. (2003). Diet choice and the role of time preference. *Journal of Consumer Affairs, 37*(1), 143–160.

Iseke, J. (2013). Indigenous storytelling as research. *International Review of Qualitative Research, 6*(4), 559–577.

Kim, M. (2002). Self-disclosure: Bragging vs. negative self-disclosure. In *Non-western perspectives on human communication: Implications for theory and practice* (pp. 121–130). SAGE Publications, Inc.

Kladnuenklum, K. (2003). *Waste recycling bank at Ruam Kan Saeng Community* (Master program in public administration). Burapha University.

Klöckner, C. A. (2013). A comprehensive model of the psychology of environmental behaviour—A meta-analysis. *Global Environmental Change, 23*(5), 1028–1038.

Kosteas, V. D. (2015). Physical activity and time preference. *International Journal of Health Economics and Management, 15*(4), 361–386.

Leenen, W. R., Grosch, H., & Groß, A. (2005). Bausteine zur interkulturellen Qualifizierung der Polizei. Waxmann Verlag.

Lim, V.-C. L., Mohd Noor, M. I., Justine, E. V., Fadzil, K. S., & Hong Ching, G. (2020). Understanding the socio-ecological systems in Tun Mustapha Park, Sabah, Malaysia through local communities' perspectives. *Paper presented at the Rufford Small Grant Conference*, Kuala Lumpur, Malaysia. https://www.rufford.org/files/Malaysia%20Conference%20 2020%20_0.pdf

Loewenstein, G. F., Weber, E. U., Hsee, C. K., & Welch, N. (2001). Risk as feelings. *Psychological Bulletin, 127*(2), 267.

Löfström, E., Klöckner, C. A., & Nesvold, I. H. (2020). Nature in your face – Disruptive climate change communication and eco-visualization as part of a garden-based learning approach involving primary school children and teachers in co-creating the future. *Frontiers in Psychology, 11*(2876), 568068.

Markus, H. R., & Kitayama, S. (2010). Cultures and selves: A cycle of mutual constitution. *Perspectives on psychological science, 5*(4), 420–430.

Markowitz, E. M., & Guckian, M. L. (2018). Climate change communication: Challenges, insights, and opportunities. In S. Clayton & C. Manning (Eds.), *Psychology and climate change* (pp. 35–63). Academic Press.

Moser, S. C. (2014). Communicating adaptation to climate change: The art and science of public engagement when climate change comes home. *Wiley Interdisciplinary Reviews: Climate Change, 5*(3), 337–358.

Newson, M., Buhrmester, M., Xygalatas, D., & Whitehouse, H. (2020). Go WILD., Not WEIRD. Journal for the Cognitive Science of Religion.

Nielsen, M., Haun, D., Kärtner, J., & Legare, C. H. (2017). The persistent sampling bias in developmental psychology: A call to action. *Journal of Experimental Child Psychology, 162*, 31–38.

Pakpour, A. H., Zeidi, I. M., Emamjomeh, M. M., Asefzadeh, S., & Pearson, H. (2014). Household waste behaviours among a community sample in Iran: An application of the theory of planned behaviour. *Waste Management, 34*(6), 980–986.

Phelan, A., Ross, H., Setianto, N. A., Fielding, K., & Pradipta, L. (2020). Ocean plastic crisis—Mental models of plastic pollution from remote Indonesian coastal communities. *PLoS One, 15*(7), e0236149.

Rad, M. S., Martingano, A. J., & Ginges, J. (2018). Toward a psychology of Homo sapiens: Making psychological science more representative of the human population. *Proceedings of the National Academy of Sciences, 115*(45), 11401–11405.

Richards, G. W., & Carruthers Den Hoed, R. (2018). Seven strategies of climate change science communication for policy change: Combining academic theory with practical evidence from science–policy partnerships in Canada. In W. Leal Filho, E. Manolas, A. M. Azul, U. M. Azeiteiro, & H. McGhie (Eds.), *Handbook of climate change communication* (Practice of climate change communication) (Vol. 2, pp. 147–160). Springer International Publishing.

Richter, I., Gabe-Thomas, E., Queirós, A. M., & Pahl, S. (n.d.-a). Maximising the potential of future scenarios by integrating psychology (in preparation).

Richter, I., Gabe-Thomas, E., Sumeldan, J., Avillanosa, A., Creencia, L., & Pahl, S. (n.d.-b). Why knowing who you are talking to is as important as your message itself. A cluster analysis on communication strategies from Palawan, Philippines Environmental Communication (in preparation).

Richter, I., Sumeldan, J., Avillanosa, A., Gabe-Thomas, E., Creencia, L., & Pahl, S. (2021). Co-Created Future Scenarios as a Tool to Communicate Sustainable Development in Coastal Communities in Palawan, Philippines. Frontiers in psychology, Section for Environmental Psychology, 21.627972, retrieved online: https://www.frontiersin.org/articles/10.3389/fpsyg.2021.627972/full

Saíd, M., Kifugo, S., Kija, H., Oiwang, G., Ojwala, M., May, R., ... Verones, F. (2019). Quantitative modelling of land use Retrieved from https://www.slideshare.net/ILRI/conservation-and-wildlife-in-kenya

Savanna Life Project. (2019). Background. Retrieved from http://www.savannalife.no/background.html

Schwartz, S. (1992). Universals in the content and structure of values: Theoretical advances and empirical tests in 20 countries. *Advances in Experimental Social Psychology, 25*(1), 1–65.

Schwartz, S. (2012). An overview of the Schwartz theory of basic values. *Online Readings in Psychology and Culture, 2*(1), 2307–0919.

Simmons, E. C., & Fielding, K. S. (2019). Psychological predictors of fishing and waste management intentions in Indonesian coastal communities. *Journal of Environmental Psychology, 65*, 101324.

Simonds, V. W., & Christopher, S. (2013). Adapting Western research methods to indigenous ways of knowing. *American Journal of Public Health, 103*(12), 2185–2192.

Singhirunnusorn, W., Donlakorn, K., & Kaewhanin, W. (2012). Contextual factors influencing household recycling behaviours: A case of waste bank project in Mahasarakham municipality. *Procedia – Social and Behavioural Sciences, 36*, 688–697.

Spence, A., Poortinga, W., & Pidgeon, N. (2012). The psychological distance of climate change. *Risk Analysis, 32*(6), 957–972.

Steg, L., & Vlek, C. (2009). Encouraging pro-environmental behaviour: An integrative review and research agenda. *Journal of Environmental Psychology, 29*(3), 309–317.

Sumeldan, J., Richter, I., Avillanosa, A., Bacosa, H., Creencia, L., & Pahl, S. (2021). Ask the Locals: A Community-Informed Analysis of Perceived Marine Environment Quality over Time in Palawan, Philippines. Frontiers in Psychology, Section for Environmental Psychology, 12:661810. Retrieved online: https://www.frontiersin.org/articles/10.3389/fpsyg.2021.661810/full

Tam, K. P., & Milfont, T. L. (2020). Towards cross-cultural environmental psychology: A state-of-the-art review and recommendations. *Journal of Environmental Psychology, 71*, 101474.

Triandis, H. C. (2001). Individualism-collectivism and personality. *Journal of Personality, 69*(6), 907–924.

Truelove, H. B., Carrico, A. R., & Thabrew, L. (2015). A socio-psychological model for analysing climate change adaptation: A case study of Sri Lankan paddy farmers. *Global Environmental Change, 31*, 85–97.

United Nations (2020). World Social Report 2020 - Inequality in a Rapidly Changing World. Retrieved online: https://www.un.org/en/file/71229/download?token=gzORmfgq

Wang, S., Hurlstone, M. J., Leviston, Z., Walker, I., & Lawrence, C. (2019). Climate change from a distance: An analysis of construal level and psychological distance from climate change. *Frontiers in Psychology, 10*, 230.

Warsono, H., Hanura, M., & Hanani, R. (2020). Governance and Livelihood of Indonesian Undocumented Migrant Workers in Sabah, Malaysia. Paper presented at the ICISPE 2019: Proceedings of the 4th international conference on Indonesian social and political enquiries, ICISPE 2019, 21–22 October 2019, Semarang, Central Java, Indonesia

Witkower, Z., & Tracy, J. L. (2020). We don't make WEIRD faces: A brief history of emotion expression research in small-scale societies. *Evolution and Human Behaviour, 41*(5), 454–455.

Youssef, J., Youssef, L., Juravle, G., & Spence, C. (2017). Plateware and slurping influence regular consumers' sensory discriminative and hedonic responses to a hot soup. *International Journal of Gastronomy and Food Science, 9*, 100–104.

Zur, I., & Klöckner, C. A. (2014). Individual motivations for limiting meat consumption. *British Food Journal, 116*(4), 629–642.

Chapter 8
Can Virtual Reality Become Real? How Immersive Virtual Experiences Might Trickle into the Real World

Amanda Elizabeth Lai and Christian A. Klöckner

8.1 Introduction

In Chap. 5 we have seen how serious games, when compared to more traditional forms of entertainment media, can promote greater cognitive learning and ability to retain knowledge in a more effective manner (*also see* Wouters et al., 2013). In this chapter, our journey exploring the integration of environmental sciences and computer graphic science will focus on the quickly emerging virtual reality technology as a tool that might be used to create disrupting events that can then fuel the further process of change, for example, in relation to climate change. Firstly, we will briefly go through some of the challenging aspects of communicating about climate change as an example of a wicked environmental problem. Then, to overcome these challenges, the chapter aims to provide a theoretical rationale for the potential of implementing immersive virtual environments (IVEs).

8.2 Communicating About Climate Change Is Challenging

If you have made it to this point in the book, you surely have a grasp of the fact that communicating about the causes, outcomes, and the risks associated to climate change and other major environmental issues is far from being an easy task, and it is more arduous than one may think (Gifford, 2011). It has been over three decades since the IPCC's first assessment report was released (Houghton et al., 1990), informing the public about the consequences of the human influence on the climate system. On top of that, climate science has matured since the 1990s, and more observations with more sophisticated models have been conducted, providing multiple lines of evidence (IPCC, 2014). However, despite the universal agreement among scientists that climate change is a compelling threat to humankind and other

C. A. Klöckner, E. Löfström, *Disruptive Environmental Communication*, Psychology and Our Planet, https://doi.org/10.1007/978-3-031-17165-9_8

life on earth (Parmesan & Yohe, 2003), public engagement toward environmental protection continues to be insufficient.

In the past decade, we have witnessed an increased media coverage of the consequences of climate change at specific occasions such as extreme weather, climate summits, or larger climate protests[1] (Hase et al., 2021), but including sustainability into mainstream discourse has apparently not taken us very far. Yet, before indulging straightaway into the fascinating potential of immersive technologies applied to the context environmental disruptive communication, we need to take a step back and find answers to questions such as the following: why do we struggle to communicate about causes and outcomes associated with climate change?

First, the dissemination of scientific results about environmental issues such as climate change still comes mostly in the form of numerical projections and descriptions to the general public that fall short in revealing the many ways in which climate change will influence our lives (Moser, 2010; Xexakis & Trutnevyte, 2021). Beyond doubt, the public understanding of the consequences of a changing climate differs substantially from the understanding of the experts, which have devoted years of education on the subject (Xexakis & Trutnevyte, 2021). Scientists experience overwhelming evidence out in the field or by analysing data that is difficult to read for the nonexperts. Accordingly, risks are then evaluated in different ways, (i.e., real risks vs. subjective judgments about risk[2]; for same examples see Siegrist & Gutscher, 2006; Lorenzoni & Hulme, 2009).

Second, research has shown that the link between public awareness about climate change and behavioral responses is not straightforward. Rising awareness and changing peoples' attitudes is not enough to generate behavior change (e.g., Blake, 2001; Kollmuss & Agyeman, 2002; Ungar, 2000). In psychology, this phenomenon is commonly known as the "attitude-behavior" gap. The reason for that people do not always act according to their standpoint is linked to the interaction of psychological, social, and environmental factors (see, e.g., Jackson, 2005; Steg et al., 2018; Stern, 2000).

8.3 How Do We Overcome Barriers to Public Engagement?

However, efficiently tailoring these difficult to read scientific findings to the public together with overcoming the potential sociopsychological barriers comes with substantial challenges. Past research has pointed at the three main components relevant to increase public engagement: (a) cognitive engagement, (b) affective engagement, and (c) behavioral engagement (Lorenzoni et al., 2007). These aspects of engagement respectively refer to the knowledge and understanding one has of the climate

[1] Even if the general trend seems to be flat across the ten analyzed countries.

[2] The perception of risk concerns individuals' subjective responses to risk and hazards; it refers to different types of attitudes regarding hazards and the risks related to them together with the way individuals judge these hazards (Breakwell, 2014).

issue, one's emotions and concern about the issue, and one's motivation to take action. Therefore, communication approaches should aim to achieve meaningful engagement in all three components (Ockwell et al., 2009).

In essence, to promote reactions in the public, climate data needs to be presented in a way that people can read, understand, and relate to (Moser, 2016). For instance, by providing feedback to the public about the impact of their actions, we can increase their awareness, and when people are aware of the ecological consequences of their lifestyles, they are more inclined to adapt their behavior to minimize their impact (Abrahamse et al., 2007; Taniguchi et al., 2003). Additionally, efficient communication strategies require an approach that comprehends how to engage the public at an affective and emotional level rather than a top-down approach based on expert understandings (Ockwell et al., 2009). A "one-way communication process," as defined by Hulme (2009), fails to meaningfully engage, because it does not consider people's underlining values, emotions, and attitudes.

Public engagement with socially relevant issues has been increasingly explored and different approaches have been put to test (see Chap. 3 for examples involving art, Chap. 5 involving games, or Chap. 6 following a provotyping approach), from designing artworks illustrating real-time energy consumption to infographics that make it easier for the public to make sense of data regarding sustainability actions or public displays and projections with the purpose to promote debates among the public (see Chap. 9).

Nevertheless, the continuous development of new technology and the emergent affordability of these technologies provided new standards to support public engagement and for communicating the causes and consequences of climate change. This also opens to a new perspective on the field of disruptive interventions. For instance, the use of information technology to communicate environmental impact as a potential solution for addressing environmental issues has grown exponentially in the last decade (Tomlinson, 2012; Shobande & Ogbeifun, 2022). From simple non-immersive information boards developed to be used to monitor people's carbon footprints, technology has evolved to immersive technologies that provide the necessary assistance for understanding complex environmental problems and exploring them actively "from the inside" (Faucheux & Nicolaï, 2011). Examples of such immersive technologies are *(i) virtual reality*, which is a digital environment that is put in place of the physical environment surrounding the user' *(ii) augmented reality*, which includes digital components that are laid over a live stream of the user's surroundings; and *(iii) mixed reality*, which includes an integration of elements of both virtual and augmented realities (Bekele et al., 2018; Danado et al., 2003; Kaplan et al., 2021).

8.4 Specifying Immersive Virtual Reality

In the next paragraphs, we will go through what immersive VR is and its potential for environmental disruptive communication.

8.4.1 What Are the Fundamental Elements of Immersive Virtual Reality?

Immersive virtual reality (IVR) is a powerful computer-generated simulation tool, which creates an effect of full *immersion* in a three-dimensional interface that simulates the user's physical presence in a *virtual world* (Flavian et al., 2018; Loureiro et al., 2018, 2020). In other words, it allows people to be immersed and interact with a virtual environment in a way that feels like a real environment. Being immersed in an alternative world refers to the sensation of "being in an environment" (Sherman & Craig, 2018) which is depending on the extent to which the technology adopted is surrounding, inclusive, vivid, and matching movements of the recipient and the system (Slater et al., 1996). Technologies used for this can be non-immersive systems, such as desktop displays, or immersive systems, such as head-mounted displays (HMDs; Pallavicini et al., 2019), the latter usually creating the more immersive experience.

As introduced in Chap. 5 in the context of gaming, immersion refers to the feeling of "being transferred" into a mediated computerized environment, in which the user's awareness of the surroundings and everyday life concerns are held back while being completely absorbed into the virtual world (Jennett et al., 2008). Additionally, immersion is related to the sense of *presence*, which is a sense of spatial immersion into the virtual environment (Steuer, 1992) and to the sense of flow, which, as explained in the Chap. 5, is the sensation of "gaming in action" in which nothing else matters (Michailidis et al., 2018; Weibel & Wissmath, 2011).

What distinguishes this computer technology from other 3D visualization systems are specific features: for example, IVR provides *sensory feedback* (i.e., realistic images, sounds, and other sensations) in response to the user input (Sherman & Craig, 2018). Realistic sensory feedback is fundamental to physical immersion, and to obtain it, key elements are stereoscopic imagery that does not exceed 1/16 second to refresh when the coordinates of the visual sensors change and position tracking. Additionally, VR requires the adoption of specific devices, such as head-mounted display (HMD), commonly known as "VR glasses." Finally, for the VR to be perceived as authentic, it needs to be *interactive* and respond to the actions of the user (Sherman & Craig, 2018). Thus, manipulation tools that give the user *agency* (i.e., the feeling of being able to exert control) over the environment, such as controllers or hand tracking systems (Nigel & Liliya, 2014), are required. All in all, a *virtual world, immersion, sensory feedback,* and *interactivity* are fundamental elements to experience VR. We will now dive deeper into the theoretical reasons that make IVEs a great tool to communicate and engage in environmental issues.

8.4.2 Can We Use Immersive Virtual Reality to Promote Engagement?

The use of IVRs is studied and informed by visual communication science (e.g., Machin, 2014). Visual experiences are an important part of our day-to-day life, and the way we encounter certain experiences can influence our behavior and our future choices. This is in line with the stimulus-organism response (S-O-R) theory, proposed by Mehrabian and Russell in the 1970s (1974),[3] which postulates that *stimuli* from the environment can trigger fundamental changes in people's cognitive and emotional state (*organism*) which in turn influences the individual's *responses*. The S-O-R theory has also provided a theoretical basis to explore the factors associated with IVEs (Kim et al., 2020). Essentially, because in VR the users think, feel, and act akin to the real world, virtual environments can accommodate for tailored experiences that are designed to activate real cognitive and emotional engagement toward environmental issues and which allow users to enact their behavioral responses in a world in which mistakes are free.

This is in line with more contemporary literature outlining the arguments for how immersive virtual environments can be implemented to influence understanding, emotions, and actions in support of climate change engagement (Beck et al., 2018):

- First (*cognitive engagement*), artificial environments can function as visualization tools (Radu, 2014), which can aid the *understanding* of complex concepts supporting people's reasoning.
- Second (*affective engagement*), artificial environments can be used to provoke **emotional responses** by manipulating the visual, auditory, and haptic stimuli (Riva et al., 2007). The latter is linked to the sense of presence that the computer-generated virtual space evokes in the user. Having the impression of being really embedded in the virtual place can support intense feelings (Sanchez-Vives & Slater, 2005). Especially this aspect is interesting from a perspective of disruptive communication as VR here can be a tool to create emotionalizing disruptive experiences in a protected virtual space.
- Last (*behavioral engagement*), IVEs might offer possibilities to promote the habituation of certain behaviors. As Beck et al. (2018) point out, interacting with the environment can boost physically embodied understandings which are complementary to the phenomena involved in climate change, supporting the formation of links between observable and unobservable phenomena.

As a matter of fact, IVEs offer experiential opportunities that can break the limits of what can be safely done in the real world, therefore opening possibilities for disruptions not possible or ethical in the real world (such as, e.g., creating the experience of a being in a flood in your own neighborhood).

---

[3] The theory has been adapted and integrated in a number of different domains across time. For example, cognition and physiology were added to the original theory by Bitner (1992).

However, are these computer-generated experiences good enough for designing interventions? We will address this question in the next section.

8.4.3 *Is Immersive Virtual Reality a Reliable Tool?*

Because in VR the users' cognition, behaviors, and experiences are analogous to those in real environments, virtual environments give us an excellent opportunity to understand and learn about human experiences by replicating and analysing complex real-life scenarios and the human experience that follow them (Kuliga et al., 2015). The use of VR as a research tool has been promoted by scholars in diverse research fields (e.g., fire evacuation research (Cao et al., 2019)), who have found that real-world experiments indeed confirm participants' perceptions and behaviors in the VR experiments. This corroborates that people's experiences are generally comparable in both VR and the real-world (Kinateder et al., 2013; Ronchi et al., 2018, 2019).

The proven external validity of these simulations allows for adopting IVR as a persuasive technology to influence and change people's attitudes and behaviors. Via simulations people can experience the effects of their actions in remarkable ways and create mental links between cause and effect, therefore initiating new ways of perceiving and knowing. In the context of environmental communication, the use of IVEs as a tool that can motivate and captivate people is increasing, as it makes it easier to learn about the impacts of human behavior on the environment, together with applying these learnings to the real world (Chirico et al., 2021; Markowitz & Bailenson, 2021; Neo et al., 2021; Queiroz et al., 2018; Scurati et al., 2021). Therefore, it seems safe to conclude that VR and IVR in particular are powerful new technologies, opening ways of environmental communication not possible before.

8.5 Conceptualizing Disruptive Communication via Immersive Virtual Reality

Still, it is an open question how IVR relates to the core of this book, disruptive communication, and in which way this confrontative mean of communicating to the public marries into the new information technology tool. In the next paragraphs, we will provide some explicative arguments for how the two may come together.

Through the lens of disruptive communication, the application of IVRs is certainly an innovative approach for challenging the current way of doing things. It is, for example, possible to digitally design immersive manifestations of the climate phenomena to confront otherwise comfortable audiences with, both, real scenarios of a nature that has been disrupted by decades of human bad choices or alternative scenarios that showcase dystopian (or utopian) pathways into the future. Therefore, by displaying the large number of stressors affecting our ecosystems in immersive scenarios and the probability of global breakdowns the current century is facing, we may be able to awoke the uneasy sense of immediacy that the disruptive communication approach aims for.

Additionally, by providing virtual experiences that do not conform with people's norms and expectations, the user is encouraged to think about the previously underrecognized meanings behind them. This process of reflection allows for criticizing one's dominant assumptions, offering a new lens through which the challenges of our time can be understood. It may be this new perspective that throws light on alternative ways to intervene effectively in the world.

Furthermore, because of their highly engaging qualities, IVRs make it possible for users to experience the effects of their actions in vivid and memorable ways, sometimes as embodied memories. This means that we can not only use IVRs to put people out of their comfort zone, which induces them to question the current status quo, but we can present the disruption as if it was real, offering a holistic experience of confrontation to the user. As well as non-immersive eco-visualizations, IVRs can stimulate creativity, but in addition, they offer the possibility to engage with the environment and to experience the outcomes of the users' actions, in the path of finding new ways of doing things. Then again, a transformational shift takes place in which **virtual reality can become real**.

Finally, IVR offers the opportunity to design disruptive experiences that are impossible to create in the real world, either because it would be dangerous or unethical to put people into them in reality or even because it is physically impossible. Imagine, for example, a IVR experience where you take the perspective of an insect and experience the threats that humanity puts on you.

Disruptive communication indeed holds the capacity to directly impact important dimensions of engagement mentioned earlier in this chapter. Providing vivid negative feedback through simulated experiences in which the user is immersed can be a potent "eye opener" that disrupts the user at the *cognitive level*, a confrontation that is logical, tangible, easy to understand, and hard to dismiss. After all we cannot "unsee" what we have seen. Additionally, simulating the damaging effects of human consumption on the planet through 3D visualizations in which the user is present may evoke strong emotions and perception of an emergency situation, thus generating disruption at the *affective level*. Likewise, direct experiences in which the users' activities in the IVE are disrupted by the simulated ecological consequences of their choices or by embodying experiences that are unfamiliar, is, by definition, disruptive at the *behavioral level*. The IVR in itself is a disruption of reality and a jump into an alternative reality. However, the IVR experiences do not necessarily need to be negative. The technology might also be used to explore (and maybe even co-create) alternative, more sustainable future pathways.

In the next segment, we showcase evidence from past research that supports these assumptions and advocates for **virtual disruptive experiences that can sustain new ways of being in the world**.

8.6 From Detached Observers to Active Participants

As mentioned earlier in this chapter, people may not always translate their attitudes into action because of the interaction of psychological, social, and environmental factors that might draw them into different directions (e.g., Jackson, 2005; Stern, 2000). One important factor for nonaction is that the climate crisis is for most people not immediately observable other than in manifestations of extreme weather events, which they might not have experienced personally. Indeed, it is still unlikely for the many individuals to experience personally a manifestation of an environmental challenge in their everyday life and have a chance of glimpsing firsthand the disruptive changes displayed by nature (although, e.g., the summer heatwaves in Europe in summer 2022 affected millions of people). The impacts of climate change (e.g., melting ice caps, ocean acidification, rising sea levels, extreme weather events, natural hazards, etc.) are not perceived directly by everybody. The difficulty to recognize that climate change is a real issue with real impacts has been linked to the fact that they are not immediately observable (Blennow et al., 2012). Although this seems to be changing with an increasing frequency and size of such events, the majority of people still perceive climate change effects indirectly, through media, news coverage, or documentaries. On top of that, most of the time, media, stakeholders, and policymakers conceptualize the environment as separate from us – although we are not a separable part of nature (Olausson, 2020). In addition, people expect the negative consequences to be far in the future rather than in the short term (Spence et al., 2012; Wolf & Moser, 2011). In essence, the fact that people do not see it and do not experience it leads to the perception of climate change as a socially, spatially, and temporally remote problem.

Since IVEs allow for communicating concepts that are less accessible, are complex, and are abstract in a more tangible and vibrant way, they carry the potential to reduce the perceived psychological distance from environmental problems (Ahn et al., 2014b; Blascovich & Bailenson, 2011). Recent studies have investigated the role that psychological distance plays in how people perceive and react to an environmental VR experience (Lorenzoni et al., 2007). Raja and Carrico (2021), for example, investigated the psychological experience of engaging with ocean acidification and its impact on coral reefs and marine life in VR with the purpose of understanding the role of psychological distance. Their VR experience consisted of four consecutive scenarios. The first scenario was located in traffic and the user could observe CO_2 molecules running out from a car's exhaust pipe. The second scenario was located on a boat where the user could observe the CO_2 molecules being absorbed by the ocean. The third scenario was located underwater where the user could explore the surroundings of a healthy coral reef environment with sea creatures. Finally, the last scenario shows a degraded rocky reef in which none of the sea creatures were present. Most of the participants described feeling psychologically close to ocean acidification after the experience, portraying a shift from being an unattached third-party witness to the phenomena (prior to VR), to feeling as if they had actually experienced it firsthand (after VR). By being presented to a nature that is twisted in a way that disrupts normal ways of thinking, the participants began to "own" the experience. Furthermore, the experience helped them understanding an unfamiliar location, and when presented with a future scenario in which coral reefs would no longer exist, they were disappointed. By being in the world differently, they started to think differently in the world, too.

This was just a simple example of the transformative power of applying IVEs as tools in disruptive communication. But the possibilities are many and the advantage is that they come with limited costs.

If people can be presented with the disturbing consequences of resources abuse projected in time, reducing the perceived temporal distance can be achievable. And by displaying catastrophic scenarios as if they were real already, the perspective of the public is allowed to change with no actual negative consequences for the user or the environment.

People often find it difficult to visualize the consequences of their actions in the present and relate them to future climate scenarios, and when they do so, they tend to frame change far from their local surroundings. Scholars have suggested that it could be a result of an evolutionary tendency to pay attention to immediate and personally relevant issues (Lorenzoni et al., 2007; Moser & Dilling, 2004). Therefore, experiencing alternative perspectives carries the potential for reshaping how people understand environmental problems that are perceived to be "far away" (Lorenzoni et al., 2007).

Adopting the point of view of others, nonetheless, being exposed to the reality of other beings may create a window of opportunity for change by reducing the dissimilarity and distance to oneself. For example, in a study by Ahn et al. (2016), participants were immersed into a virtual environment in which they embodied a cow. They were asked to get down on their hands and knees, and they experienced being loaded onto a lorry and being hit[4] by a cattle prod. The results of this study showed that embodying sensory-rich experiences of the cattle led to higher interconnection between the self and nature which in turn provoked stronger perceptions of imminence of the environmental risk.

In another experiment conducted by Ahn et al. (2014a), participants experienced cutting down a virtual tree. In the VR, participants were immersed in a forest, surrounded by trees and the sound of birds chirping. The participants were asked to hold the handle of a chain saw which was synced to a haptic joystick and were instructed to move back and forth to cut down a virtual tree. After 2 minutes, the tree trunk crashed down to the ground. After cutting the tree, the birds stopped chirping, and the forest became quiet to call attention to the destruction caused to the forest. Ultimately, the virtual experience led to a 20% reduction in usage of paper napkins among the users compared to a control condition in which the participants had read a pamphlet describing the cutting of a tree.

---

[4]The experience of being hit was executed with the haptic stimuli of being poked in the back with the combination of sound and vibration.

Displaying the disturbing consequences of their action in the VR symbolizes the disruptive component that allowed participants to reset the way they made decisions about their behavior in the real world. The disruptive communication in this case happened at the same point in time in which the paper napkins were consumed. It could be argued that the experience might have linked the participants' behavior closer to their values. The findings of the abovementioned studies point to the fact that IVEs can be a powerful tool in reducing psychological distance and in closing the knowledge-to-action gap.

8.7 Immersive Virtual Disaster Experiences

In addition to the disruptive qualities discussed in the previous section, IVRs can be a valuable asset for analyzing human behavior during crisis. Having the means to produce realistic scenarios of hazards, with IVRs it is possible to elicit realistic feelings and study the users' behaviors in scenarios that would be too dangerous and costly to implement in real-world crises, such as earthquakes (Li et al., 2017) and fires (Kinateder et al., 2014). Furthermore, immersive technologies can be adopted for communicating real-world climate risks to evoke adequate coping responses.

Risk analysis literature shows that direct disaster experiences are effective in making people aware of their own vulnerability to future climate change-related disasters (Bronfman et al., 2020) and direct experience prompts sharper judgments regarding the effectiveness of coping responses when dealing with risks (Zaalberg & Midden, 2010). The affective dimension is particularly important in determining risk perception. Based on the risk analysis literature, we can conclude that negative emotions can influence the way we feel about a given risk by increasing our reactions to danger (Slovic, 2000).

Most of the research focusing on risk communication has been conducted, though, by framing messages which are conveyed by traditional means of communication (Slovic, 2000). Potential developments **in conveying risks** consist in replacing framed messages with framed virtual experiences. The use of 3D representations in virtual and augmented realities is an emerging field of research in the literature for environmental risk communication (for examples see Bailey et al., 2015; Jerald, 2015). When communicating climate change adaptive strategies, policymakers, and environmental scientists may use 3D visualization techniques to understand and communicate environmental risks.

Since urban flooding is of growing concern, let us take the case of flood risk management as an example; a critical part of risk management is risk communication which seeks to prepare the population at risk for an emergency situation (Argent et al., 2016; Maskrey et al., 2016). Risk literature shows that having direct flooding experiences is one of the strongest factors to prompt awareness of one's own vulnerability and understanding which coping strategies are effective (Burningham et al., 2008; Lechowska, 2018; Mondino et al., 2020; Spence et al., 2011).

However, many have never experienced a flood directly and are not aware of the potential impacts (Bradford et al., 2012). Additionally, nonexpert audiences can find it difficult to understand traditional flood hazard maps because they do not correspond to the user needs (e.g., Rollason et al., 2018; Hagemeier-Klose & Wagner, 2009). As a matter of fact, the probabilistic risk information of a cartography can be hard to interpret (Dransch et al., 2010; O'Sullivan et al., 2012; Strathie et al., 2017) which leads to flood maps not being helpful at all for mitigation activities (Leskens et al., 2014). So how do we get the attention of people who have not been affected so far?

IVEs can make disaster data understandable to the lay person and support their coping responses to flood risks (Qiu et al., 2017; Zaalberg & Midden, 2013). It is possible to create experiences that enhance people's understanding of these emergency situations without putting their lives at risk and improve their awareness of the consequences of responding inappropriately. When immersed in the simulation, they can experience firsthand whether the flood will hit their proprieties and quickly understand if the water will be limited to the ground floor or if it will reach the first floor, or, for instance, make a judgment if it is safer to evacuate to the street or moving up to the roof. Furthermore, the vivid and realistic imagery can support estimating the potential damages (Liu et al., 2017). This is especially relevant for individuals living in flood-prone areas but do not have direct flooding experiences. Actively living an experience while being immersed in a highly realistic virtual environment (vs. passively receiving an external message) can start a transformative process in which the creation of new guiding principles occurs.

For example, in a study by Zaalberg and Midden (2010), participants were exposed to an interactive 3D simulation of a dike breach which flooded their virtual residence. In their findings, the authors report that the participants' motivation to evacuate from the virtual threat had increased. But even more interestingly, the experience had increased the participants' motivation to buy flood insurance in the real world. Feeling "physically" present in high-fidelity representations of flood inundations is certainly a powerful tool to engage the public and stakeholders and improve flood resilience. IVEs are effective substitutes for personal experience providing significant contribution in hazard communication and risk perception.

Yet, we are far from fully harnessing these techniques as they require the application and integration of knowledge from several disciplines (Samuels, 2011).

8.8 Environmental Connectedness in IVR

Finally, we would like to draw attention to the potential for VR technology to create positively disrupting experiences. The use of virtual environments can also promote the feeling of environmental connectedness, so in a way disrupting everyday life in a positive direction, by creating beautiful virtual experiences of nature.

Environmental connectedness has been defined a fundamental belief in which peoples' attitudes toward environmental issues are rooted (Bruni et al., 2008; Geng et al., 2015; Janpol & Dilts, 2016). Research has shown that people that have a stronger connection to nature will most likely behave in a more sustainable way (Mayer & Frantz, 2004; Nisbet et al., 2009; Perkins, 2010; Schultz, 2000, 2002). The key determinant of the relationship between people and nature is how much one perceives nature to be part of oneself, which strengthens the likelihood of environmental behaviors (Ahn et al., 2016; Davis et al., 2009; Soliman et al., 2017; Wang et al., 2019; Zelenski et al., 2015).

Considering that the sense of feeling connected to nature is directly influenced by the frequency of one's contact with nature since childhood (Thompson et al., 2008), scholars have found that experiences with nature are the key for effective interventions (e.g., Cleary et al., 2018; Ernst & Theimer, 2011). However, having the opportunity to have face to face contact with nature is a prerogative of a more privileged public (White et al., 2018). Thereupon, a new line of research has been growing in popularity, which investigates the effects of mediated nature experiences simulated by computer technologies on environmental connectedness (Breves & Heber, 2020). Not all mediated experiences can evoke compelling feelings of connectedness with nature, yet it has been proven that immersive virtual environments can elicit realistic nature experiences that affect people in the same way as being physically in direct contact with nature (Kisker et al., 2021; Schöne et al., 2019; Segovia & Bailenson, 2009), therefore creating feelings of connectedness between humans and nature (Ahn et al., 2016; Breves & Heber, 2020). Of course, the immersive qualities of the virtual environment need to be significantly good in imitating nature to allow the experience to be superior to regular media in eliciting nature connectedness (Klein & Hilbig, 2018; Zelenski et al., 2015).

8.9 Summary and Concluding Remarks

In this chapter, we have made a case for that immersive virtual realities can be a very interesting and potent case of disruptive communication. They allow for creating experiences that are out of the ordinary (hence, potentially disruptive), even outside what is physically possible in the real world. They can give people the opportunity to take perspectives of other human beings, of animals, and even of plants or objects. This experience might disrupt people's thinking substantially. VR allows experiencing events that are real and happening but have not happened yet for a respective person (like a flood induced by climate change-related extreme precipitation). VR can even create positively disrupting experiences that stimulate connectedness to nature in new ways. The act of immersing into a virtual reality is a disruption in itself as it stops people from being in their reality for some time. Thus, VR and IVR are promising tools of disruption that offer a lot of potential for the future. However, VR also has its pitfall, starting from it inducing motion sickness in some people (Kim et al., 2018), still limited technical options to interact with the virtual world, and the fact that people are usually very aware that the experience is a simulation, which reduces the transferability to the real world to some degree (an effect that is also seen for games in environmental communication; see Chap. 5).

References

Abrahamse, W., Steg, L., Vlek, C., & Rothengatter, T. (2007). The effect of tailored information, goal setting, and tailored feedback on household energy use, energy-related behaviors, and behavioral antecedents. *Journal of Environmental Psychology, 27*, 265–276. https://doi.org/10.1016/j.jenvp.2007.08.002

Ahn, S. J., Bailenson, J. N., & Park, D. (2014a). Short- and long-term effects of embodied experiences in immersive virtual environments on environmental locus of control and behavior. *Computers in Human Behavior, 39*, 235–245. https://doi.org/10.1016/j.chb.2014.07.025

Ahn, S. J., Fox, J., Dale, K. R., & Avant, J. A. (2014b). Framing virtual experiences: Effects on environmental efficacy and behavior over time. *Communication Research, 42*(6), 839–863.

Ahn, S. H., Bostick, J., Ogle, E., Nowak, K. L., McGillicuddy, K. T., & Bailenson, J. N. (2016). Experiencing nature: Embodying animals in immersive virtual environments increases inclusion of nature in self and involvement with nature. *Journal of Computer-Mediated Communication, 21*(6), 399–419. https://doi.org/10.1111/jcc4.12173

Argent, R., Sojda, R., Giupponi, C., Mcintosh, B. S., Voinov, A., & Maier, H. (2016). Best practices for conceptual modelling in environmental planning and management. *Environmental Modelling & Software, 80*, 113–121. https://doi.org/10.1016/j.envsoft.2016.02.023

Bailey, J. O., Bailenson, J. N., Flora, J., Armel, K. C., Voelker, D., & Reeves, B. (2015). The impact of vivid messages on reducing energy consumption related to hot water use. *Environment and Behavior, 47*(5), 570–592.

Beck, D., Allison, C., Morgado, L., Pena-Rios, A., Ogle, T., Richter, J., Gütl, C., & Pirker, J.. (2018). *ILRN 2018 Montana*. https://doi.org/10.3217/978-3-85125-609-3

Bekele, M. K., Pierdicca, R., Frontoni, E., Malinverni, E. S., & Gain, J. (2018). A survey of augmented, virtual, and mixed reality for cultural heritage. *Journal on Computing and Cultural Heritage, 11*(2)., 7:1–7:36. https://doi.org/10.1145/3145534

Bitner, M. J. (1992). Servicescapes: The impact of physical surroundings on customers and employees. *Journal of Marketing, 56*(2), 57–71. https://doi.org/10.2307/1252042

Blake, D. (2001). Contextual effects on environmental attitudes and behavior. *Environment and Behavior, 33*, 708–725. https://doi.org/10.1177/00139160121973205

Blascovich, J., & Bailenson, J. N. (2011). Infinite reality: Avatars, eternal life, new worlds, and the dawn of the virtual revolution. In NA: Vol. NA (Issue NA, p. NA).

Blennow, K., Persson, J., Tomé, M., & Hanewinkel, M. (2012). Climate change: Believing and seeing implies adapting. *PLoS One, 7*(11), e50182. https://doi.org/10.1371/journal.pone.0050182

Bradford, R. A., O'Sullivan, J. J., van der Craats, I. M., Krywkow, J., Rotko, P., Aaltonen, J., Bonaiuto, M., De Dominicis, S., Waylen, K., & Schelfaut, K. (2012). Risk perception – Issues for flood management in Europe. *Natural Hazards and Earth System Sciences, 12*(7), 2299–2309. https://doi.org/10.5194/nhess-12-2299-2012

Breakwell, G. M. (2014). *The psychology of risk* (2nd ed., p. xiv, 368). Cambridge University Press.

Breves, P., & Heber, V. (2020). Into the wild: The effects of 360° immersive nature videos on feelings of commitment to the environment. *Environmental Communication, 14*(3), 332–346. https://doi.org/10.1080/17524032.2019.1665566

Bronfman, N. C., Cisternas, P. C., Repetto, P. B., Castañeda, J. V., & Guic, E. (2020). Understanding the relationship between direct experience and risk perception of natural hazards. *Risk Analysis, 40*(10), 2057–2070.

Bruni, C. M., Fraser, J., & Schultz, P. W. (2008). The value of zoo experiences for connecting people with nature. *Visitor Studies, 11*(2), 139–150. https://doi.org/10.1080/10645570802355489

Burningham, K., Fielding, J., & Thrush, D. (2008). 'It'll never happen to me': Understanding public awareness of local flood risk. *Disasters, 32*(2), 216–238. https://doi.org/10.1111/j.1467-7717.2007.01036.x

Cao, L., Lin, J., & Li, N. (2019). A virtual reality based study of indoor fire evacuation after active or passive spatial exploration. *Computers in Human Behavior, 90*, 37–45.

Chirico, A., Scurati, G. W., Maffi, C., Huang, S., Graziosi, S., Ferrise, F., & Gaggioli, A. (2021). Designing virtual environments for attitudes and behavioral change in plastic consumption: A comparison between concrete and numerical information. *Virtual Reality, 25*(1), 107–121. https://doi.org/10.1007/s10055-020-00442-w

Cleary, A. F., Kelly, S., Murray, Z., & Roiko, A. (2018). Predictors of nature connection among urban residents: Assessing the role of childhood and adult nature experiences. *Environment and Behavior, 52*(6), 001391651881143–001391651881610. https://doi.org/10.1177/0013916518811431

Danado, J., Dias, E., Romao, T., Correia, N., Trabuco, A., Santos, C., Araujo, D., Duarte, P., Rebocho, R., Palmeiro, J., Serpa, J., Costa, M., & Camara, A. (2003). *Mobile Augmented Reality for Environmental Management (MARE)*. https://doi.org/10.2312/egs.20031075

Davis, J. L., Green, J. D., & Reed, A. (2009). Interdependence with the environment: Commitment, interconnectedness, and environmental behavior. *Journal of Environmental Psychology, 29*(2), 173–180. https://doi.org/10.1016/j.jenvp.2008.11.001

Dransch, D., Rotzoll, H., & Poser, K. (2010). The contribution of maps to the challenges of risk communication to the public. *International Journal of Digital Earth, 3*(3), 292–311. https://doi.org/10.1080/17538941003774668

Ernst, J., & Theimer, S. (2011). Evaluating the effects of environmental education programming on connectedness to nature. *Environmental Education Research, 17*(5), 577–598. https://doi.org/10.1080/13504622.2011.565119

Faucheux, S., & Nicolaï, I. (2011). IT for green and green IT: A proposed typology of eco-innovation. *Ecological Economics, 70*(11), 2020–2027. https://ideas.repec.org/a/eee/ecolec/v70y2011i11p2020-2027.html

Flavian, C., Ibáñez Sánchez, S., & Orús, C. (2018). The impact of virtual, augmented and mixed reality technologies on the customer experience. *Journal of Business Research*. https://doi.org/10.1016/j.jbusres.2018.10.050

Geng, L., Xu, J., Ye, L., Zhou, W., & Zhou, K. (2015). Connections with nature and environmental behaviors. *PLoS One, 10*(5), e0127247. https://doi.org/10.1371/journal.pone.0127247

Gifford, R. (2011). The dragons of inaction: Psychological barriers that limit climate change mitigation and adaptation. *American Psychologist, 66*(4), 290.

Hagemeier-Klose, M., & Wagner, K. (2009). Evaluation of flood hazard maps in print and web mapping services as information tools in flood risk communication. *Natural Hazards and Earth System Sciences, 9*(2), 563–574. https://doi.org/10.5194/nhess-9-563-2009

Hase, V., Mahl, D., Schäfer, M. S., & Keller, T. R. (2021). Climate change in news media across the globe: An automated analysis of issue attention and themes in climate change coverage in 10 countries (2006–2018). *Global Environmental Change, 70*, 102353.

Houghton, J. T., Jenkins, G. J., & Ephraums, J. J. (Eds.). (1990). *Climate change: The IPCC scientific assessment*. Cambridge University Press.

Hulme, M. (2009). *Why we disagree about climate change: Understanding controversy, inaction and opportunity—Why we disagree about climate change: Understanding controversy, inaction and opportunity*. In NA: Vol. NA (Issue NA, p. NA). https://doi.org/10.1017/cbo9780511841200

IPCC (2014). *Climate change 2014: Synthesis report. Contribution of working groups I, II and III to the fifth assessment report of the Intergovernmental Panel on Climate Change* [Core Writing Team, R.K. Pachauri and L.A. Meyer (eds.)]. IPCC: Geneva, Switzerland.

Jackson, T. (2005). Motivating sustainable consumption. *Sustainable Development Research Network, 29*(1), 30–40.

Janpol, H. L., & Dilts, R. (2016). Does viewing documentary films affect environmental perceptions and behaviors? *Applied Environmental Education & Communication, 15*(1), 90–98. https://doi.org/10.1080/1533015X.2016.1142197

Jennett, C. C., Anna, L., Cairns, P., Dhoparee, S., Epps, A., Tijs, T., & Walton, A. (2008). Measuring and defining the experience of immersion in games. *International Journal of Human-Computer Studies, 66*(9), 641–661. https://doi.org/10.1016/j.ijhcs.2008.04.004

Jerald, J. (2015). *The VR book: Human-centered design for virtual reality.* Morgan & Claypool.

Kaplan, A. D., Cruit, J., Endsley, M., Beers, S. M., Sawyer, B. D., & Hancock, P. A. (2021). The effects of virtual reality, augmented reality, and mixed reality as training enhancement methods: A meta-analysis. *Human Factors, 63*(4), 706–726. https://doi.org/10.1177/0018720820904229

Kim, H. K., Park, J., Choi, Y., & Choe, M. (2018). Virtual reality sickness questionnaire (VRSQ): Motion sickness measurement index in a virtual reality environment. *Applied Ergonomics, 69*, 66–73.

Kim, M. J., Lee, C.-K., & Jung, T. (2020). Exploring consumer behavior in virtual reality tourism using an extended stimulus-organism-response model. *Journal of Travel Research, 59*(1), 69–89.

Kinateder, M., Pauli, P., Müller, M., Krieger, J., Heimbecher, F., Rönnau, I., Bergerhausen, U., Vollmann, G., Vogt, P., & Mühlberger, A. (2013). Human behaviour in severe tunnel accidents: Effects of information and behavioural training. *Transportation Research Part F: Traffic Psychology and Behaviour, 17*, 20–32. https://doi.org/10.1016/j.trf.2012.09.001

Kinateder, M., Ronchi, E., Nilsson, D., Kobes, M., Müller, M., Pauli, P., & Mühlberger, A. (2014). *Virtual reality for fire evacuation research* (pp. 313–321). https://doi.org/10.15439/2014F94

Kisker, J., Gruber, T., & Schöne, B. (2021). Experiences in virtual reality entail different processes of retrieval as opposed to conventional laboratory settings: A study on human memory. *Current Psychology, 40*. https://doi.org/10.1007/s12144-019-00257-2

Klein, S. A., & Hilbig, B. E. (2018). How virtual nature experiences can promote pro-environmental behavior. *Journal of Environmental Psychology, 60*, 41–47. https://doi.org/10.1016/j.jenvp.2018.10.001

Kollmuss, A., & Agyeman, J. (2002). Mind the gap: Why do people act environmentally and what are the barriers to pro-environmental behavior? *Environmental Education Research, 8*(3), 239–260. https://doi.org/10.1080/13504620220145401

Kuliga, S. F., Thrash, T., Dalton, R. C., & Hölscher, C. (2015). Virtual reality as an empirical research tool—Exploring user experience in a real building and a corresponding virtual model. *Computers, Environment and Urban Systems, 54*, 363–375.

Lechowska, E. (2018). What determines flood risk perception? A review of factors of flood risk perception and relations between its basic elements. *Natural Hazards, 94*(3), 1341–1366. https://doi.org/10.1007/s11069-018-3480-z

Leskens, J. G., Brugnach, M., Hoekstra, A. Y., & Schuurmans, W. (2014). Why are decisions in flood disaster management so poorly supported by information from flood models? *Environmental Modelling & Software, 53*, 53–61. https://doi.org/10.1016/j.envsoft.2013.11.003

Li, C., Liang, W., Quigley, C., Zhao, Y., & Yu, L.-F. (2017). Earthquake safety training through virtual drills. *IEEE Transactions on Visualization and Computer Graphics, 23*(4), 1275–1284. https://doi.org/10.1109/TVCG.2017.2656958

Liu, J., Gong, J.-H., Liang, J.-M., Li, Y., Kang, L.-C., Song, L.-L., & Shi, S.-X. (2017). A quantitative method for storm surge vulnerability assessment – A case study of Weihai city. *International Journal of Digital Earth, 10*(5), 539–559. https://doi.org/10.1080/17538947.2016.1229052

Lorenzoni, I., & Hulme, M. (2009). Believing is seeing: laypeople's views of future socio-economic and climate change in England and in Italy. *Public Understanding of Science, 18*(4), 383–400.

Lorenzoni, I., Nicholson-Cole, S., & Whitmarsh, L. (2007). Barriers perceived to engaging with climate change among the UK public and their policy implications. *Global Environmental Change, 17*(3–4), 445–459. https://doi.org/10.1016/j.gloenvcha.2007.01.004

Loureiro, S., Guerreiro, J., Eloy, S., Langaro, D., & Panchapakesan, P. (2018). Understanding the use of Virtual Reality in Marketing: A text mining-based review. *Journal of Business Research*. https://doi.org/10.1016/j.jbusres.2018.10.055

Loureiro, S., Romero, J., & Bilro, R. G. (2020). Stakeholder engagement in co-creation processes for innovation: A systematic literature review and case study. *Journal of Business Research, 119*(NA), 388–409. https://doi.org/10.1016/j.jbusres.2019.09.038

Machin, D. (Ed.). (2014). *Visual communication* (Vol. 4). Walter de Gruyter.

Markowitz, D. M., & Bailenson, J. N. (2021). Virtual reality and the psychology of climate change. *Current Opinion in Psychology, 42*, 60–65. https://doi.org/10.1016/j.copsyc.2021.03.009

Maskrey, S. A., Mount, N. J., Thorne, C. R., & Dryden, I. (2016). Participatory modelling for stakeholder involvement in the development of flood risk management intervention options. *Environmental Modelling & Software, 82*, 275–294. https://doi.org/10.1016/j.envsoft.2016.04.027

Mayer, F. S., & Frantz, C. M. (2004). The connectedness to nature scale: A measure of individuals' feeling in community with nature. *Journal of Environmental Psychology, 24*(4), 503–515. https://doi.org/10.1016/j.jenvp.2004.10.001

Mehrabian, A., & Russell, J. A. (1974). *An approach to environmental psychology*. The MIT Press.

Michailidis, L. B.-B., & Emili; He, Xun. (2018). Flow and immersion in video games: The aftermath of a conceptual challenge. *Frontiers in Psychology, 9*(NA), 1682–1682. https://doi.org/10.3389/fpsyg.2018.01682

Mondino, E., Scolobig, A., Borga, M., & Di Baldassarre, G. (2020). The role of experience and different sources of knowledge in shaping flood risk awareness. *Water, 12*(8), 2130. https://doi.org/10.3390/w12082130

Moser, S. C. (2010). Communicating climate change: History, challenges, process and future directions. *Wiley Interdisciplinary Reviews: Climate Change, 1*(1), 31–53.

Moser, S. C. (2016). Reflections on climate change communication research and practice in the second decade of the 21st century: What more is there to say? *Wiley Interdisciplinary Reviews: Climate Change, 7*(3), 345–369.

Moser, S. C., & Dilling, L. (2004). Making climate HOT. *Environment: Science and Policy for Sustainable Development, 46*(10), 32–46. https://doi.org/10.1080/00139150409605820

Neo, J. R. J., Won, A. S., & Shepley, M. M. (2021). Designing immersive virtual environments for human behavior research. *Frontiers in Virtual Reality, 2*. https://www.frontiersin.org/article/10.3389/frvir.2021.603750

Nigel, F., & Liliya, K. (2014). Past and future applications of 3-D (virtual reality) technology. *Научно-Технический Вестник Информационных Технологий, Механики и Оптики, 6*(94), 1–8.

Nisbet, E. K., Zelenski, J. M., & Murphy, S. A. (2009). The nature relatedness scale: Linking individuals' connection with nature to environmental concern and behavior. *Environment and Behavior, 41*(5), 715–740. https://doi.org/10.1177/0013916508318748

O'Sullivan, J. J., Bradford, R. A., Bonaiuto, M., De Dominicis, S., Rotko, P., Aaltonen, J., Waylen, K., & Langan, S. J. (2012). Enhancing flood resilience through improved risk communications. *Natural Hazards and Earth System Sciences, 12*(7), 2271–2282. https://doi.org/10.5194/nhess-12-2271-2012

Ockwell, D., Whitmarsh, L., & O'Neill, S. (2009). Reorienting climate change communication for effective mitigation: Forcing people to be green or fostering grass-roots engagement? *Science Communication, 30*(3), 305–327.

Olausson, U. (2020). Making sense of the human-nature relationship: A reception study of the "Nature Is Speaking" campaign on YouTube. *Nature and Culture, 15*(3), 272–295.

Pallavicini, F. P., & Alessandro; Minissi, Maria Eleonora. (2019). Gaming in virtual reality: What changes in terms of usability, emotional response and sense of presence compared to non-immersive video games? *Simulation & Gaming, 50*(2), 136–159. https://doi.org/10.1177/1046878119831420

Parmesan, C., & Yohe, G. (2003). A globally coherent fingerprint of climate change impacts across natural systems. *Nature, 421*(6918), 37–42. https://doi.org/10.1038/nature01286

Perkins, H. E. (2010). Measuring love and care for nature. *Journal of Environmental Psychology, 30*(4), 455–463. https://doi.org/10.1016/j.jenvp.2010.05.004

Qiu, L., Du, Z., Zhu, Q., & Fan, Y. (2017). An integrated flood management system based on linking environmental models and disaster-related data. *Environmental Modelling & Software: With Environment Data News, 91*, 111–126. https://doi.org/10.1016/j.envsoft.2017.01.025

Queiroz, A. C. M., Kamarainen, A. M., Preston, N. D., & da Silva Leme, M. I. (2018). Immersive virtual environments and climate change engagement. *ILRN 2018 Montana*, 153.

Radu, I. (2014). Augmented reality in education: A meta-review and cross-media analysis. *Personal and Ubiquitous Computing, 18*(6), 1533–1543.

Raja, U. S., & Carrico, A. R. (2021). A qualitative exploration of individual experiences of environmental virtual reality through the lens of psychological distance. *Environmental Communication, 15*(5), 594–609. https://doi.org/10.1080/17524032.2020.1871052

Riva, G., Mantovani, F., Capideville, C. S., Preziosa, A., Morganti, F., Villani, D., Gaggioli, A., Botella, C., & Alcañiz, M. (2007). Affective interactions using virtual reality: The link between presence and emotions. *Cyberpsychology & Behavior, 10*(1), 45–56.

Rollason, E., Bracken, L. J., Hardy, R. J., & Large, A. R. G. (2018). Rethinking flood risk communication. *Natural Hazards, 92*(3), 1665–1686. https://doi.org/10.1007/s11069-018-3273-4

Ronchi, E., Fridolf, K., Frantzich, H., Nilsson, D., Walter, A. L., & Modig, H. (2018). A tunnel evacuation experiment on movement speed and exit choice in smoke. *Fire Safety Journal, 97*, 126–136. https://doi.org/10.1016/j.firesaf.2017.06.002

Ronchi, E., Mayorga, D., Lovreglio, R., Wahlqvist, J., & Nilsson, D. (2019). Mobile-powered head-mounted displays versus cave automatic virtual environment experiments for evacuation research. *Computer Animation and Virtual Worlds, 30*(6). https://doi.org/10.1002/cav.1873

Samuels, P. (2011). Development of good professional practice. *Journal of Flood Risk Management, 4*, 1–2. https://doi.org/10.1111/j.1753-318X.2010.01089.x

Sanchez-Vives, M. V., & Slater, M. (2005). From presence to consciousness through virtual reality. *Nature Reviews Neuroscience, 6*(4), 332–339.

Schöne, B., Wessels, M., & Gruber, T. (2019). Experiences in virtual reality: A window to autobiographical memory. *Current Psychology, 106*, 1–5. https://doi.org/10.1007/s12144-017-9648-y

Schultz, P. W. (2000). New environmental theories: Empathizing with nature: The effects of perspective taking on concern for environmental issues. *Journal of Social Issues, 56*(3), 391–406. https://doi.org/10.1111/0022-4537.00174

Schultz, P. W. (2002). Inclusion with nature: The psychology of human-nature relations. In P. Schmuck & W. P. Schultz (Eds.), *Psychology of sustainable development* (pp. 61–78). Springer. https://doi.org/10.1007/978-1-4615-0995-0_4

Scurati, G. W., Bertoni, M., Graziosi, S., & Ferrise, F. (2021). Exploring the use of virtual reality to support environmentally sustainable behavior: A framework to design experiences. *Sustainability, 13*(2), 943. https://doi.org/10.3390/su13020943

Segovia, K. Y., & Bailenson, J. N. (2009). Virtually true: Children's acquisition of false memories in virtual reality. *Media Psychology, 12*(4), 371–393. https://doi.org/10.1080/15213260903287267

Sherman, W. R., & Craig, A. B. (2018). *Understanding virtual reality: Interface, application, and design*. Morgan Kaufmann.

Shobande, O. A., & Ogbeifun, L. (2022). Has information and communication technology improved environmental quality in the OECD?—A dynamic panel analysis. *International Journal of Sustainable Development & World Ecology, 29*(1), 39–49.

Siegrist, M., & Gutscher, H. (2006). Flooding risks: A comparison of lay people's perceptions and expert's assessments in Switzerland. *Risk Analysis, 26*(4), 971–979.

Slater, M., Linakis, V., Usoh, M., & Kooper, R. (1996). Immersion, presence and performance in virtual environments: An experiment with tri-dimensional chess. *Proceedings of the ACM Symposium on Virtual Reality Software and Technology*, 163–172. https://doi.org/10.1145/3304181.3304216

Slovic, P. E. (2000). *The perception of risk*. Earthscan Publications.

Soliman, M., Peetz, J., & Davydenko, M. (2017). The impact of immersive technology on nature relatedness and pro-environmental behavior. *Journal of Media Psychology: Theories, Methods, and Applications, 29*(1), 8–17. https://doi.org/10.1027/1864-1105/a000213

Spence, A., Poortinga, W., Butler, C., & Pidgeon, N. F. (2011). Perceptions of climate change and willingness to save energy related to flood experience. *Nature Climate Change, 1*(1), 46–49. https://doi.org/10.1038/nclimate1059

Spence, A., Poortinga, W., & Pidgeon, N. (2012). The psychological distance of climate change. *Risk Analysis, 32*(6), 957–972. https://doi.org/10.1111/j.1539-6924.2011.01695.x

Steg, L., van den Berg, A. E., & de Groot, J. I. M. (2018). Environmental psychology. In *Environmental psychology* (pp. 1–11). John Wiley & Sons. https://doi.org/10.1002/9781119241072.ch1

Stern, P. C. (2000). New environmental theories: Toward a coherent theory of environmentally significant behavior. *Journal of Social Issues, 56*(3), 407–424. https://doi.org/10.1111/0022-4537.00175

Steuer, J. (1992). Defining virtual reality: Dimensions determining telepresence. *Journal of Communication, 42*(4), 73–93. https://doi.org/10.1111/j.1460-2466.1992.tb00812.x

Strathie, A., Netto, G., Walker, G. H., & Pender, G. (2017). How presentation format affects the interpretation of probabilistic flood risk information. *Journal of Flood Risk Management, 10*(1), 87–96. https://doi.org/10.1111/jfr3.12152

Taniguchi, A., Hara, F., Takano, S., Kagaya, S., & Fujii, S. (2003). Psychological and behavioral effects of travel feedback program for travel behavior modification. *Transportation Research Record, 1839*(1), 182–190.

Thompson, C. W., Aspinall, P., & Montarzino, A. (2008). The childhood factor: Adult visits to green places and the significance of childhood experience. *Environment and Behavior, 40*(1), 111–143. https://doi.org/10.1177/0013916507300119

Tomlinson, B. (2012). *Greening through IT: Information technology for environmental sustainability*. MIT Press.

Ungar, S. (2000). Knowledge, ignorance and the popular culture: Climate change versus the ozone hole. *Public Understanding of Science, 9*(3), 297–312. https://doi.org/10.1088/0963-6625/9/3/306

Wang, J., Geng, L., Schultz, P. W., & Zhou, K. (2019). Mindfulness increases the belief in climate change: The mediating role of connectedness with nature. *Environment and Behavior, 51*(1), 3–23. https://doi.org/10.1177/0013916517738036

Weibel, D., & Wissmath, B. (2011). Immersion in computer games: The role of spatial presence and flow. *International Journal of Computer Games Technology, 2011*(NA), 6–14. https://doi.org/10.1155/2011/282345

White, M. P., Yeo, N. L., Vassiljev, P., Lundstedt, R., Wallergård, M., Albin, M., & Lõhmus, M. (2018). A prescription for "nature" – The potential of using virtual nature in therapeutics. *Neuropsychiatric Disease and Treatment, 14*, 3001–3013. https://doi.org/10.2147/NDT.S179038

Wolf, J., & Moser, S. C. (2011). Individual understandings, perceptions, and engagement with climate change: Insights from in-depth studies across the world. *WIREs Climate Change, 2*(4), 547–569. https://doi.org/10.1002/wcc.120

Wouters, P., van Nimwegen, C., van Nimwegen, C., van Oostendorp, H., & van der Spek, E. D. (2013). A meta-analysis of the cognitive and motivational effects of serious games. *Journal of Educational Psychology, 105*(2), 249–265. https://doi.org/10.1037/a0031311

Xexakis, G., & Trutnevyte, E. (2021). Empirical testing of the visualizations of climate change mitigation scenarios with citizens: A comparison among Germany, Poland, and France. *Global Environmental Change, 70*, 102324.

Zaalberg, R., & Midden, C. (2010). PERSUASIVE – Enhancing human responses to climate change risks through simulated flooding experiences. *Persuasive Technology, 6137*, 205–210. https://doi.org/10.1007/978-3-642-13226-1_21

Zaalberg, R., & Midden, C. J. H. (2013). Living behind dikes: Mimicking flooding experiences. *Risk Analysis, 33*(5), 866–876. https://doi.org/10.1111/j.1539-6924.2012.01868.x

Zelenski, J. M., Dopko, R. L., & Capaldi, C. A. (2015). Cooperation is in our nature: Nature exposure may promote cooperative and environmentally sustainable behavior. *Journal of Environmental Psychology, 42*, 24–31. https://doi.org/10.1016/j.jenvp.2015.01.005

Chapter 9
Two Disruptive Eco-visualizations: Initial Experiences

Erica Löfström, Kristoffer S. Fjællingsdal, Isabel Richter, and Christian A. Klöckner

9.1 Introduction

Throughout the previous chapters, we presented the concept of disruptive communication, the opportunities and challenges that come with it, and how games, artwork, and future scenarios can become disruptive communication by themselves. If you are anything like us, you are probably bursting with disruptive communication ideas and concepts by now, concepts that you cannot wait to explore further. We would like to finish this book by introducing you to two disruptive eco-visualizations that are currently trialled as part of our most recent project called Nature in Your Face (NIYF),[1] generously funded by the Norwegian Research Council (grant no. 302111). The project is based on the very idea of using disruptive eco-visualizations to communicate climate-related issues to citizens and stimulate real action.

NIYF is an ongoing 4-year project aiming to generate system change by disruption. Currently, we are in the process of developing disruptive eco-visualizations for four parallel case studies in Norway on different topical areas (mobility, housing, food waste, and plastic neutrality). These disruptive eco-visualizations will be tested and evaluated with actual users. So far, we have developed two such eco-visualization concepts – the "Lung Tree" for the mobility case and the "Food Box" for the food waste case – and built fully functioning prototypes for both concepts. Actual user testing and evaluations are being carried out in 2022. In the following paragraphs, we would like to share some initial experiences from working with these two disruptive eco-visualization prototypes with you. First, we will briefly describe the two concepts, explain the theoretical background, and illustrate the resulting prototypes. We then share our experiences from initial user tests with each of them. We hope that by sharing these experiences, we will prepare you for what might happen in the

[1] www.niyf.no

C. A. Klöckner, E. Löfström, *Disruptive Environmental Communication*, Psychology and Our Planet, https://doi.org/10.1007/978-3-031-17165-9_9

real world should you want to explore the possibilities of disruptive communication within your field of expertise.

9.2 The Lung Tree: Breathing or Not Breathing – A Life Support Machine

The initial "Lung Tree" visualization idea was developed by Erica Löfström. Subsequently, it was elaborated further in an applied master-level interdisciplinary university course at the Norwegian University of Science and Technology which is referred to as "Experts in Teams" (https://www.ntnu.no/eit/psy3809). A student group used an early version of the "Nature in Your Face" project idea and developed the concept of the Lung Tree idea further.[2] Combining the Lung Tree with other eco-visualization ideas, the concept of a holistic "eco-visualization-walkway" with different eco-visualization stations, which were all connected to aspects of modern urban mobility, was developed.[3]

The visual similarities between an X-ray of a human lung and the branches of trees inspired the idea of a tree that breathes using audio and visual elements (sounds of breathing and pulsating red light illuminating the central branches of the tree). This makes it appear as a "Breathing Lung Tree," in other words, a tree where the interaction with air becomes visible and hearable to the citizens passing by. The breathing sound and light would be synchronized to give an organic feel to the tree, making the tree look and sound like an inflating and deflating lung. This lung would then be connected to real-time air quality data from the space the tree would be installed. When air quality is good, the tree will breathe with a healthy and pure sound and light movement; when air quality is poor, it would be more strained with a higher breathing frequency and more heavy sounds of breathing (with high air pollution, even the sound of fighting for air). This initial concept was developed further together with the artist Stanza[4] who was an artist in residence at NTNU at that time, and the Art and Technology Task Force (ARTEC)[5] at NTNU, coordinated by Andrew Perkis. As a result, an audiovisual portable art installation which was given the name "Breathing or not breathing: A-life support machine" was built (see Fig. 9.1).[6] Unfortunately, the intention to use local air quality sensor data to feed the visualization had to be abandoned for technical reasons, and the visualization instead uses global sensor data.

---

[2] We thank Andrew Perkis, Dagfinn Dahl Dybvig, Sara Brinch, Jordi Puig, NTNU ARTEC and STANZA for their invaluable contribution to realizing the idea.

[3] Thank you to Anima Bista Thapa, Eirik Vige Kristoffersen, Jacob Qvam Skavang, and Hedvig Tønnesen for the permission to work further on their ideas.

[4] https://stanza.co.uk

[5] https://www.ntnu.edu/artec

[6] https://stanza.co.uk/breathing/index.html

Fig. 9.1 The Lung Tree installed in Trondheim, Norway

The artist Stanza describes the visualization like this:

> Using a custom-made API [Application Programming Interface – a software that allows communication between two applications] that gathers data from 120 world cities collecting NO_2 data [nitrogen dioxide – a harmful gas], the system is overseen by an AI [Artificial Intelligence] which uses machine learning to detect patterns that help the tree to breathe. What you experience is a time-based audio-visual artwork in a tree. (https://stanza.co.uk/breathing/index.html)

As part of the NIYF project, we installed the Lung Tree for a public test run in the city center of Trondheim in Winter 2022. A poster was attached to the tree to inform visitors about the project. This poster features information about the background of the installation and a QR code to allow people to respond to the visualization via a short online questionnaire. In addition, the reactions of the public were assessed via observations and spontaneous dialogues with people who passed by the tree. Below, we share some of our initial experiences and responses from the public.

The Lung Tree was planned to "breathe" all day round for 2 weeks. However, almost immediately, we received complaints from people living close by as the breathing sound was disturbing them in their night sleep. As a consequence, a timer was installed, turning the tree off automatically between 9 PM and 8 AM. As the complaints persisted, we had to further reduce the "breathing time" to 1 hour a day, between 8 and 9 PM. One obvious conclusion about disruptive audio- and visual eco-visualizations in public spaces is that their installation needs to be carefully communicated to anyone living in its direct vicinity to give people the possibility to raise their voices ahead of time. In the case of the Lung Tree, the sound level was comparatively low compared to the sounds of rush hour traffic, but it seems that a combination of the unfamiliarity of the sound as well as low traffic noise levels during the night still disturbed the neighbors. The installation was definitely disruptive, as the complaints clearly demonstrated. However, it was maybe disruptive in a different way than intended.

Of course, it was not our intent to disturb peoples' daily lives to the point where our interventions have negative impacts on health and wellbeing. What we realized is that there is a clear distinction between disturbing and disrupting. Disturbing people could either be (1) disturbing them in terms of interrupting their sleep schedules or being needlessly invasive in their daily lives or (2) emotionally disturbing, as in how the tree sounded kind of disgusting. At least the first understanding of disturbing is what we did not want to achieve. Disruptive is more about bringing people out of the confines of their daily lives and comfort zones and providing them with an experience (positive or negative) that could, at least in theory, make them think a little harder about their actions and lifestyles. We argue that the tree was both disturbing and disruptive but also think that the impact it had on peoples' peace and quiet was the deciding factor for it being turned off preemptively. The main intention behind the tree was, of course, to make people stop and ponder the installation – not for neighboring individuals to lose sleep over it. Interestingly, one of the complaining neighbors contacted us again after we had put the timer on and said that he thought the idea and message behind the installation were good; he just needed his night's sleep.

For future use of audiovisual eco-visualizations, it might be an option to place the installation in public areas with lots of passersby but few permanent residents, such as large public parks. Another idea is to hold separate events where the Lung Tree is demonstrated in combination with workshops to channel the reactions onto creative solution-finding processes, as was outlined in Chap. 1 of this book. For further implementations of the Lung Tree, these follow-up workshops are planned anyway as part of the Nature in Your Face concept, but in this initial test run, the workshops were not implemented, yet. As such, the installation could likely be suitable for conference arenas or hotels featuring indoor nature installations.

On the bright side, we did get not only complaints but also a lot of positive responses to the Lung Tree. We did notice that a lot of people who were passing by the tree stopped and talked about it, clearly indicating its disruptive potential. It was placed in an area of Trondheim where – during the evening hours when the tree was active – many people pass by on the way to or from bars and restaurants. Many of these people were clearly amused and laughed as they went by. Others looked puzzled and tried to figure out what it was about. We clearly managed to make many people stop in their activities and reflect. As such, the installation also achieved its goal to be disruptive in a positive sense, as described in Chap. 1.

On one occasion, a little girl and her mother passed by the tree, and we saw them talking about it as they went by. One of the researchers approached them:

Hello, I saw you looking at the tree, what do you think?

The mother nodded to her daughter, as encouraging her to speak, and the girl replied:

It's weird with a breathing tree in the middle of the city… She then appeared thoughtful and added:

But then, all trees are breathing, aren't they? A middle-aged lady spontaneously exclaimed:

It's the planet breathing, isn't it?

Many of the people passing by were wondering about the meaning of the different breathing sounds with respect to air pollution data. We assume that the connection between the actual sensor data and the visualization needs to be communicated more clearly in the future. In addition, the use of local data instead of aggregated global data would have made it more relevant to its visitors, a notion that is also underlined by researchers on climate change visualization who say that climate change should rather be visualized by relating it to local effects (Sheppard, 2012).

Other visitors interpreted the Lung Tree as a delayed, dystopian Christmas tree, Valentine's Day installation or advertisement and thereby neglected the relationship to air pollution and climate change. This shows that using a visual language established by other actors (here, the red of the lights which triggered Valentine's Day associations and the tree illumination used in shopping areas) may have its downfalls. However, it can also be used in a positive way by slightly altering established visual expressions and thereby causing the intended disruption. At least for the ones thinking of a dystopian Christmas tree, this aspect seems to have worked.

Our initial assessments showed that the Lung Tree managed to evoke emotional responses such as surprise and a sense of awe. This is reflected both in expressions collected from people on-site and preliminary analyses of the first survey responses. The preliminary, descriptive results from the survey ($N = 16$) indicate that people mainly react with surprise. The majority of respondents agree that the artwork is something novel, something they have never seen before, and that it has a confrontational undertone. All these aspects are characteristics of artworks that might trigger a disruption. However, when it comes to behavioral consequences, most respondents indicate that they would not like to change their own behavior after having visited the Lung Tree, which is also in line with the concept introduced in Chap. 1, as the follow-up workshop, which should channel the initial activation into action was not implemented, yet.

The responses on site, however, were of a more general character and did not specifically revolve around air quality as such. However, it did seem to trigger thoughts on our human relation to the ecosystem as a whole, which is quite interesting. Another intriguing, albeit perhaps slightly discouraging finding is that art installations like the Lung Tree often tend to fail in communicating their purpose to a general population. Art is, by nature, subjective. As such, there is no real guarantee that a single installation or artistic vision is easily understood or communicated. It is, for example, likely that the "globality" of the Lung Tree – gathering climate data from all over the world rather than the municipality in which it was stationed – could have lessened its impact on some of the viewers. Additionally, it is worth addressing that art installations such as the Lung Tree are affected by interference from the artist that makes them, researchers utilizing them for educational or persuasive purposes, and the audience that utilizes them for leisure, entertainment, or objects of fascination and rumination. A climate-minded artist would, for instance, perhaps be more likely to emphasize the communicative properties of the pro-environmental elements and aspects in their art piece than an artist who is more interested in social issues. Similarly, scientists exhibiting and doing research on climate art are likely to determine the relative success or failure of an art installation depending on the data they gather about it. For example, if an art installation is designed to visualize changes in air quality, a scientist might deem it a failed experiment if their respondents did not exhibit any increased understanding or thoughts about air quality after viewing it or interacting with it. Lastly, the diverse audience that views the installation is likely to perceive it in accordance with their own attitudes, morals, values, and cognitive schemas. Altogether, the interplay between artist, scientist, and audience are important aspects to consider for future studies. Altogether, however, we are excited to explore the responses to the Lung Tree further and to make additional user studies on its effects and potential as a discussion starter.

9.3 Food Box: Food (Waste) in Your Face

The other disruptive eco-visualization we have developed so far as part of NIYF is an installation we call the "Food Box." This eco-visualization started as an idea among our case study partners in the food waste case.[7] For this NIYF case study, we were planning to carry out case studies with four Folk High Schools in Norwegian communities. At these schools, all pupils live at the school with all meals being served at the local school's canteen. NIYF collaborates with the schools to minimize their food waste.

For this, an eco-visualization has been co-developed by project team members: the Food Box is a portable installation based on live-video technology that presents its users with their own food waste at the moment the food waste is disposed of by the pupils (see Fig. 9.2). As such, it is aimed at interrupting the process of food disposal when the pupils are cleaning their plates from non-eaten food in the food waste bin. The installation can be easily transported with a small van or car and does not take up much space at respective school canteens.

At an early stage of co-development of the concept, the food waste left on the plate was turned into a beautiful kaleidoscope picture on the screen in the Food Box. However, the project team considered the resulting beautiful pictures are rather rewarding than disrupting. Therefore, during five internal workshops, this initial idea was gradually transformed into a more disruptive concept, namely, a digital visual fusion of the content of the plate and the users' faces so that their faces appeared to be covered by their own food waste – in essence, "Food Waste in Your Face." This solution was the result of intense discussion within the group around the desired level of disruptiveness of this intervention. Finding the right level of disruptiveness was something that also challenged the project group, and the project

[7] We thank Elise Matilde Malik, Astrid Fuglesang Stokke, Knut Ankjell, Ken Lid, Skjalg Molvær, Herman Tandberg, Ann Kristin Andreassen, Mona Nilsen, Torstein Maugesten, and Hilde Rønning for their invaluable contribution to realizing the idea.

Fig. 9.2 The Food Box installed in a school in Norway

leaders had to go back to the initial idea of disruptive communication and thoroughly discuss the necessity of this aspect before the discussions eventually landed on the final concept. In short, the disruptiveness of the visualization seemed to challenge those in the group who were more used to and comfortable with less disruptive communication strategies.

After the first two test runs, it became clear that the Food Box had to be adapted further: not only pupils who have leftover food but also pupils who do not have any food waste need some form of feedback, something that had not been considered before. This underlines the importance of co-creation and the relevance of user feedback. One idea could be to portray the pupil's clean face surrounded by a halo or shining like the sun. We are currently working on developing the concept further, integrating feedback we received from pupils and teachers.

The visualization itself seemed to amuse the pupils, and many of them took a selfie of their own food-face while in the Food Box as well as with us after the experience. We also encouraged the pupils to share their Food in Your Face selfie and other reflections on the Food Box on social media using the hashtag #NIYF. We wish to develop this feature as part of the project and follow-up with additional studies on how this may be utilized to create additional engagement for the theme visualized on a larger scale. Some teachers were also lining up to test the Food Box during our demonstration. Apparently, the Food Box was a clearly disruptive element in the school canteens, interfering with the usual fast, efficient, and not reflected process of throwing away uneaten (but edible) food remaining on the plates. Whereas this usually is a process with high automaticity, as it is conducted every day in a very stable context (Klöckner & Verplanken, 2012), with the intervention of the Food Box, this has become a process creating a lot of attention and discussion.

From the first experiences with the vision workshops following the Food Box experience, we also get the impression that the Food Box can be a promising tool for generating creative, often realistic suggestions regarding the minimization of food waste from school canteens. We are still analyzing a questionnaire survey from both schools, but the creative solutions that were presented by the pupils as part of the vision workshops speak their own clear language. These solutions range from changing giving visual feedback about produced food waste (e.g., transparent waste disposal bins so that one might see the food waste more clearly, or putting up some form of weight in which one can see how much food waste is being generated) to interventions specifically targeting factors leading to food waste more generally (e.g., encouraging the kitchen staff to make food that suits the pupils' tastes, not needlessly mixing ingredients before serving them, or allowing the pupils themselves to decide how much food to put on their plates). Furthermore, it was suggested to allow taking "doggie bags" of food back to their rooms, organizing team competitions on who throws away the least edible food, getting a piglet that could be fed with the food waste and eventually eating the pig. Another idea was to use the food waste to generate compost, which would become the basis of a school garden where again new food could be grown. All suggestions were gathered and discussed in the plenum together with the pupils, and the pupils were invited to prioritize among these suggestions. We are currently reviewing all suggestions and preparing a presentation to the schools, who will then follow-up on some of the suggestions as the project continues.

One problematic aspect that is still unsolved is that due to the technology behind the Food Box, each participant spends 30–45 seconds on average inside the box in order to carry out the intervention and have their picture taken. This resulted in a significant lineup in the canteen, disrupting the efficiency of the processes. However, as the intervention was temporary, this disruption might even be regarded as an opportunity, as it creates a space for dialogue and reflection. The negative effect of waiting too long can be reduced by having more than one Food Box installed.

The Food Box technology can be programmed to visualize other things. For instance, we have discussed the possibility of letting the box visualize the visitor's own face merged with ocean plastic waste that they may have picked up from the shore or the very plastic packaging of the food that they may have just bought at the local grocery store. Also, we have an ongoing project with a Norwegian natural science museum that is planning an exhibition on local endangered species and the problem of invasive alien species taking over the local ecosystem. So far, the idea is to let the museum visitors take a quiz about what they learned at the exhibition, and if they fail the test, they will appear to get an invasive species "invade" their face, while if they pass the test, they will have local species surrounding their clean face.

9.4 Conclusions

Looking back at the theoretical frameworks presented earlier in this book, the experiences with the two disruptive elements outlined in this chapter can teach us important lessons. From our initial experiences, it appears that the Food Box stimulated very clear and concrete ideas for how to tackle the issue of food waste in a very specific situation, the school canteen. There is a very clear link between the problematic behavior, the visualization of it (involving both the actor – the pupil – and the point in time – the disposal), and the vision workshops, which were conducted in close temporal proximity to the experience. For the Lung Tree, the necessary line of thought between the experience, the own contribution, and its solution is longer and less guided by the installation. This apparently reduced the effect of the Lung Tree installation in comparison to the Food Box. This resonates well with earlier experiences made with a combination of disruptive communication elements and vision workshops (Löfström & Klöckner, 2019; Löfström & Klöckner, 2020; Löfström et al., 2021), where also an application to a relatively specific topic (the plastic waste in oceans) resulted in promising results from the process. A lesson learned from our early applications can therefore be to make sure that the topics addressed by disruptive communication should not be too abstract.

The two examples of disruptive communication further show that they seem to address at least some of the barriers to change outlined in Chap. 2: the Food Box clearly interferes with the habitual behavior of disposing of food waste, which is an opportunity for reflection and creating new, in this case, more sustainable habits (Verplanken & Roy, 2016). The Lung Tree also interfered with a potentially habitualized part of behavior, namely, commuting to and from a shopping, bar, and restaurant area. It succeeded in making people stop and read the poster. Both installations also clearly had a component focusing on social norms and values (Cialdini & Jacobson, 2021), as both were implemented in situations where people from the target group were together with peers. Clearly, both installations were able to start discussions between these people, a good start for reflecting on what acceptable norms are in the relevant social groups. Whether the installations appealed to factors of identity is less clear. It can be assumed that the mapping of food waste on the face of the pupils makes the link to their personal or group identity (Fritsche et al., 2018; Whitmarsh & O'Neill, 2010) salient, but we have no clear evidence for that from the first experiences. Even less clear it is for the Lung Tree. With regard to pushing people further in the stages of change, as outlined in Chap. 2 (Bamberg, 2013), we did not test this explicitly in the first implementations. However, we conclude that the Lung Tree installation itself (without a vision workshop) seems not sufficient to get people closer to changing their behavior, which is also in line with experiences from the Climart project (Roosen, 2020). When the disruptive element is combined with a vision workshop, as in the Food Box example, though, it appears that the process of behavior change at least proceeds into the planning stage.

By sharing our very first experiences of using disruptive eco-visualizations to communicate sustainability and climate change-related issues with you, we hope that we have managed to convey both the vast possibilities of this exciting approach but also to help you stay clear of some of the potential pitfalls. Disruptive communication is by no means a safe and predictable path, but we do believe it represents a very promising and potentially powerful tool that may be used as part of the solution toward creating our sustainable future.

References

Bamberg, S. (2013). Changing environmentally harmful behaviors: A stage model of self-regulated behavioral change. *Journal of Environmental Psychology, 34*, 151–159.

Cialdini, R. B., & Jacobson, R. P. (2021). Influences of social norms on climate change-related behaviors. *Current Opinion in Behavioral Sciences, 42*, 1–8.

Fritsche, I., Barth, M., Jugert, P., Masson, T., & Reese, G. (2018). A social identity model of pro-environmental action (SIMPEA). *Psychological Review, 125*(2), 245.

Klöckner, C. A., & Verplanken, B. (2012). Yesterday's habits preventing change for tomorrow? About the influence of automaticity on environmental behaviour. In L. Steg, A. E. van den Berg, & J. I. M. De Groot (Eds.), *Environmental psychology: An introduction* (pp. 197–209). Wiley-Blackwell.

Löfström, E., & Klöckner, C. A. (2019). *Nature in your face: Framing co-creative visioning*. Paper presented at the ECEEE summer school, Hyeres, France.

Löfström, E., & Klöckner, C. A. (2020). Nature in your face-disruptive climate change communication and eco-visualization as part of a garden-based learning approach involving primary school children and teachers in co-creating the future. *Frontiers in Psychology, 11*, 2876.

Löfström, E., Richter, I., & Nesvold, I. H. (2021). Disruptive communication as a means to engage children in solving environmental challenges: A case study on plastic pollution. *Frontiers in Psychology, 3599.*

Roosen, L. J. (2020). *Environmental art for environmental engagement-can visual art change the way we think & feel about global environmental issues?* (PhD). Norwegian University of Science and Technology, Trondheim.

Sheppard, S. R. J. (2012). *Visualizing Climate Change*. Routledge.

Verplanken, B., & Roy, D. (2016). Empowering interventions to promote sustainable lifestyles: Testing the habit discontinuity hypothesis in a field experiment. *Journal of Environmental Psychology, 45*, 127–134.

Whitmarsh, L., & O'Neill, S. (2010). Green identity, green living? The role of pro-environmental self-identity in determining consistency across diverse pro-environmental behaviours. *Journal of Environmental Psychology, 30*(3), 305–314.

Correction to: Disruptive Environmental Communication

Christian A. Klöckner and Erica Löfström

Correction to:
C. A. Klöckner, E. Löfström, *Disruptive Environmental Communication*, Psychology and Our Planet, https://doi.org/10.1007/978-3-031-17165-9

The book was inadvertently published with an incorrect author name Erika Löfström instead of Erica Löfström. It has been corrected and updated in the book now.

We had missed to include the chapter author names in the Table of Contents of Front matter and in the Chapter Opening Pages of all the chapters. It has now been updated in the book.

The updated original version of this book can be found at:
https://doi.org/10.1007/978-3-031-17165-9

C. A. Klöckner, E. Löfström, *Disruptive Environmental Communication*,
Psychology and Our Planet, https://doi.org/10.1007/978-3-031-17165-9_10

Index

C. A. Klöckner, E. Löfström, *Disruptive Environmental Communication*, Psychology and Our Planet, https://doi.org/10.1007/978-3-031-17165-9

The manufacturer's authorised representative in the EU is Springer
Nature Customer Service Centre GmbH, Europaplatz 3, 69115 Heidelberg,
Germany. If you have any concerns regarding our products, please
contact ProductSafety@springernature.com

Printed and bound by CPI Group (UK) Ltd, Croydon, CR0 4YY
28/09/2026
02224071-0018